Praise for *The Regenerative Agriculture Solution*

"Every citizen should read this book because without the production of food and fiber from the world's land and waters, we cannot have a church, economy, political party, city, or any business. Agriculture is our lifeblood. Until the public is better-informed, current industrial agriculture, based on chemistry and technology instead of biological science, will continue. Such agriculture is the most environmentally, socially, and economically destructive ever known and will ensure continued desertification and climate change. A more informed public can lead to political parties not putting their financing from corporate profit above human welfare."

—Allan Savory, president, Savory Institute; chairman, Africa Centre for Holistic Management

"For decades both Ronnie Cummins and André Leu have been leading advocates for global food systems to be natural, healthy, and ecologically sustainable. The words in *The Regenerative Agriculture Solution* emerge from their lives of action and service. It is hard to accept that Ronnie is no longer with us, but it is a joy to learn from this new iteration of his wisdom and experience. Great appreciation, too, is owed to André's dedication to ensuring that his friend and colleague's legacy continues to educate and inspire."

—John D. Liu, ecosystem ambassador, Commonland Foundation

"This book is a testament to the vision of the late Ronnie Cummins. His friend André Leu memorializes Cummins's lifelong work with this overview of the demonstrable benefits of regenerative agriculture for everything in the book's subtitle and more. Cummins's case study on agave illustrates these benefits perfectly, making this book a useful as well as touching tribute."

—Marion Nestle, professor emerita at NYU; author of *Slow Cooked: An Unexpected Life in Food Politics*

"In *The Regenerative Agriculture Solution*, Ronnie Cummins offers a final gift: a dynamic model for using CAM (Crassulacean acid metabolism) plants to regenerate land, feed livestock, pull down carbon, and revive communities in hot, water-stressed areas. Cummins tells the story of the Billion Agave Project, beginning with the enterprising brothers who devised a way to ferment the thick, spiky agave leaves for animal fodder, saving their farm in the process. Regeneration International's André Leu situates the project within a growing wave of regional eco-restoration efforts—and shares how embracing and applying these efforts can bring us a long way toward ameliorating climate change. This is the book that Ronnie, an 'incorrigible optimist' and champion of farmers, would have wanted."

—**Judith Schwartz**, author of *The Reindeer Chronicles*

"Ronnie Cummins's work late in life at Vía Orgánica Ranch demonstrated that even in the driest of valleys in Mexico, farmers and herders can develop climate-resilient food production strategies that need not tax the soil, the aquifer, or the rural community, but enhance them. In my mind and heart, he is up there with Bill Mollison and Masanobu Fukuoka for shining a bright, warm light into our future on Planet Desert."

—**Gary Paul Nabhan**, James Beard Award–winning writer; desert agroecologist

"*The Regenerative Agriculture Solution* is a book of hope, inspiration, and solid ideas for how we might focus on the principles of building soil health. If these principles are applied to even a small percent of the Earth, we start to heal our planet and reverse the climate chaos that surrounds us. This book is worth reading and sharing, and its ideas are worth implementing."

—**Bob Quinn**, president, Quinn Farm & Ranch; founder, Kamut International; director, Quinn Institute

The Regenerative Agriculture Solution

A Revolutionary Approach to Building Soil, Creating Climate Resilience, and Supporting Human and Planetary Health

Ronnie Cummins *and* André Leu

Foreword by Vandana Shiva

Chelsea Green Publishing
White River Junction, Vermont
London, UK

Project Manager: Rebecca Springer
Developmental Editor: Brianne Goodspeed
Copy Editor: Will Solomon
Proofreader: Fran Pulver
Indexer: Shana Milkie
Designer: Melissa Jacobson
Page Layout: Abrah Griggs

Printed in Canada.
First printing September 2024.
10 9 8 7 6 5 4 3 2 1 24 25 26 27 28

Our Commitment to Green Publishing

Chelsea Green sees publishing as a tool for cultural change and ecological stewardship. We strive to align our book manufacturing practices with our editorial mission and to reduce the impact of our business enterprise in the environment. We print our books using vegetable-based inks whenever possible. This book may cost slightly more because it was printed on paper that contains recycled fiber, and we hope you'll agree that it's worth it. *The Regenerative Agriculture Solution* was printed on paper supplied by Marquis that is made of recycled materials and other controlled sources.

Library of Congress Cataloging-in-Publication Data

Names: Cummins, Ronnie, author. | Leu, Andre, author. | Shiva, Vandana, author of foreword.
Title: The regenerative agriculture solution : a revolutionary approach to building soil, creating climate resilience, and supporting human and planetary health / Ronnie Cummins and André Leu ; foreword by Vandana Shiva
Description: White River Junction, Vermont : Chelsea Green Publishing, 2024 | Includes bibliographical references and index
Identifiers: LCCN 2024024952 (print) | LCCN 2024024953 (ebook) | ISBN 9781645022695 (paperback) | ISBN 9781645022701 (ebook)
Subjects: LCSH: Organic farming. | Crops and climate. | Forest resilience—Climatic factors. | Agaves.
Classification: LCC S605.5 C86 2024 (print) | LCC S605.5 (ebook) | DDC 631.5/84—dc23/eng/20240715
LC record available at https://lccn.loc.gov/2024024952
LC ebook record available at https://lccn.loc.gov/2024024953

Chelsea Green Publishing
White River Junction, Vermont, USA
London, UK
www.chelseagreen.com

Contents

Foreword

"Never doubt that a small group of thoughtful,
committed citizens can change the world.
Indeed, it is the only thing that ever has."
—**Margaret Mead**

I am happy to be writing this foreword to a book about regenerative agriculture—not only because I believe that regenerative agriculture offers a solution to so many of the world's problems, but also because this particular book was started by my friend and longtime collaborator, Ronnie Cummins, and completed by my other friend and longtime collaborator, André Leu, after Ronnie passed away in April 2023. The book also came into being thanks in no small part to the efforts of another friend, Rose Welch, Ronnie's widow and cofounder (with him) of the Organic Consumers Association (OCA). Like the book itself, this foreword is a tribute to Ronnie—and his decades-long dedication to organic agriculture, environmental activism, and real, healthy food—as well as a celebration of the long friendship we all enjoyed. It is also a call to carry on the work.

"Regeneration" was not a commonly used word prior to 2014, when a handful of activists—me, Ronnie, and André, as well as Dr. Hans Herren from the Millennium Institute and

Steve Rye from Mercola—met in New York at the Rodale headquarters to find ways to strengthen our movement and deepen the convergence between ecological sustainability and social justice in times of climate havoc. The following year, we gathered again, this time at a biodynamic farm in Costa Rica, and formally launched Regeneration International. Ronnie was instrumental in making it happen.

I'd known Ronnie since 1988, when we met in Washington, DC, at a meeting on climate change. We were together in Rome in 1996 at the UN Food Summit, where we dreamed up an annual "fortnight of action," which would take place every year from October 2nd (Gandhi's birthday) to October 16th, World Food Day. The idea was to keep the movement energized by making it an annual event. As of 2024, the tradition continues.

At the 2015 UN Climate Change Conference in Paris, we planted a Garden of Hope and made a "Pact with the Planet" to regenerate the Earth. It was then that we also cooked up the idea of staging an international tribunal at the Hague to try Monsanto for its crimes against the Earth and its people. Ronnie and André again played a major role. Ronnie also founded the Millions Against Monsanto Campaign.

Through the years, we continued to work together, united by a passion for truth, a commitment to care, and the urge to cooperate and share our knowledge and resources. We were also united as farmers, in farming and caring for the land through regeneration. Our scientific efforts have always rested on a foundation of practical activities—they grow from doing. Knowledge and action are a continuum, flowing from the living systems and processes of the living Earth.

As André summarizes here in these pages:

"Regeneration International asserts that to heal our planet, all agricultural systems should be regenerative, organic, and based on the science of agroecology. Farmers can determine acceptable and regenerative practices using . . . [the] Four Principles of Organic Agriculture. These principles are:

1. **Health.** Organic agriculture should sustain and enhance the health of soil, plant, animal, human, and the planet as one and indivisible.
2. **Ecology.** Organic agriculture should be based on living ecological systems and cycles, work with them, emulate them, and help sustain them.
3. **Fairness.** Organic agriculture should build on relationships that ensure fairness in the familiar environment and life opportunities.
4. **Care.** Organic agriculture should be managed in a precautionary and responsible manner to protect the health and well-being of current and future generations and the environment."

This book, *The Regenerative Agriculture Solution*, distills the knowledge gained over decades of collective work—and shows the many interconnections of living systems. As André puts it, "Plants, through photosynthesis, use solar energy to turn carbon dioxide and water into glucose. Glucose is the basis of the food system for most of life. It is a primary energy source of the cells of most living organisms, including plants and animals. It is the basis of the fuels that power the mitochondria—the engines inside the cells of nearly every organism on Earth, including us."

This perspective helps us see through false solutions to climate change, such as carbon offsets, geoengineering, biofuels, and renewable energy based on the mining of lithium and cobalt. It also clarifies the real solutions that find their foundational principles in the biosphere, in living soil, and in justice and fairness among people and all living organisms.

Understanding this distinction is essential to the future health of the planet and its inhabitants. For example, carbon offsets allow polluters to continue polluting while gaining access to the resources of non-polluters. The more honest, fair, and ecological solution is "the polluter pays." As André argues, "Instead of trading carbon as a commodity on financial markets, polluters should be required to remunerate the people who are actually doing the work of regenerating the climate and environment."

It is in this spirit—a commitment to honoring and rewarding those who are doing the difficult work of regenerating the planet—that this book is published. It identifies the degenerative practices of corporate industrial agriculture trying to cloak itself as "regenerative" while also describing the principles of real regeneration, drawn from how living systems work, grow, and heal. Regenerative agriculture is an ecological solution to climate change. The book shows how regenerating our forests, rangelands, and farming ecosystems can cool our planet, restore the climate, and enrich our communities.

It is a book of hope based in knowledge and practice. Read it. Practice the principles of regeneration. Grow life. Grow hope.

—Dr. Vandana Shiva

Preface

Ronnie Cummins started this book. It was initially intended to be a short booklet on the multiple benefits of the agave agroforestry system. He submitted a book proposal and a draft chapter to his publisher, Chelsea Green. The response was that they needed it to be a more comprehensive overview of regenerative systems and not just confined to agaves.

Ronnie developed a new proposal with summaries of the new chapters. He received a favorable response and started writing this book. Ronnie and I (André) were close friends and were in regular contact. I was one of the people he regularly contacted for technical advice for the expanded proposal, and as he began to write this book.

Tragically, he passed away quite suddenly in April 2023, after finishing only two draft chapters. Rose Welch, Ronnie's widow, and Ben Trollinger, the editor at Chelsea Green, reached out to me to ask if I would consider finishing the book since I was the person who best knew what Ronnie was writing.

Naturally, I agreed, as Ronnie would have wanted me to do this.

Two chapters, chapter 1 and chapter 6, were written by Ronnie, and are his original text with only minor editing for typos, so his voice and personality are maintained.

I have written the remainder of the chapters based on our numerous discussions and plans over many years. Chapter 7 is based on a program that Ronnie and I developed as part of a significant project for Regeneration International to scale up payments for agroecological, regenerative, and organic ecosystem services. Chapters 2, 3, 4, and 5 are entirely in my voice and I have written all the text.

Consequently, the book, having two authors writing separate chapters, maintains two voices. Please think of this as a discussion between two friends who are in close agreement.

This book is a shared joint vision and, very importantly, a roadmap on how we can regenerate our people and planet.

CHAPTER ONE

Epiphany in the Desert

June 28, 2019. Membrillo, Mexico. We pull into the parking lot at the Vía Orgánica farm-to-table restaurant. It's a little after noon and the sun is already beating down. With a few clouds passing overhead, we're all praying for an end to the dry season. We've had almost no rain since October. But this is not unusual in the drylands. Sixty to seventy percent of Mexico's terrain is classified as arid, or semiarid desert, typically with no rain whatsoever for eight to nine months a year. Past the front gate, bumping down the driveway, a dust cloud rising behind, *la pipa*, the water truck, is bringing us enough water to supply the farm for a day or two.

Eighty-six percent of Mexican farmers do not have a well, and are unlikely to ever have one. In the arid and semiarid drylands, deep irrigation wells are very expensive, water rights are hard to come by, and many or most underground aquifers are overexploited. This makes it very difficult to provide forage for your livestock during the dry season. Purchasing hay and alfalfa is expensive, especially given the low-to-nonexistent market for livestock products produced by subsistence farmers. Many traditional dryland farmers are still planting corn at the beginning of the rainy season, in June or July, but without irrigation, and today's droughts and unpredictable rainfall mean that most years you don't even get mature *mazorcas* or corncobs. Your harvest usually consists of not much more than corn stalks or *rastrojo*, which you can feed to your animals in the dry season. This amounts to a lot of work and expense for what is ultimately a low-quality feed supplement.

Typically, only large landowners or agro-export farms have wells for irrigation. The rest of us have to make do with whatever falls from the sky. In addition to the high costs of buying water rights and drilling down 650 feet or more to hit water, most aquifers in our regional *La Independencia* watershed are depleted and/or contaminated with heavy metals.

A decade ago, at our 75-acre Vía Orgánica Agroecology Center and research farm, we used to pride ourselves on maintaining our greenhouses and gardens, pastures, and wooded nature refuge area without a well. We did this by capturing up to 3 million gallons of rainfall annually from the roofs of our buildings as well as storing water runoff from the nearby mountains in our ponds and cisterns during the summer rainy season. All of our buildings (handmade natural adobe with tile roofs and metal gutters to catch the rainfall) have cisterns in the basement. But nowadays, after expanding our market gardens and *milpa* (corn, beans, and squash) and starting to reforest our land, we're spending several thousand dollars a week during the dry season on water truck deliveries just to hang on. Geographers and climate scientists recently have described the extended weather forecast, from the southwestern United States to northern Mexico, as a "thousand-year drought."

Vía Orgánica's olive trees, pomegranates, mulberry, and cover crops are thirsty. We're producing a variety of healthy organic vegetables, herbs, and flowers every week, using drip irrigation from our cisterns, and our free-range poultry, sheep, and goats, as well as our pigs and rabbits are getting by; but the costs of organic grain (for the poultry and pigs) and alfalfa and hay (for the other animals) during the long dry season are unsustainable. I certainly understand now why many of the farmers in our area have largely abandoned their corn fields or *milpas*. Their main agricultural pursuit is to put their routinely malnourished animals out to fend for themselves on the sparse vegetation of their shared (eroded and badly overgrazed) communal rangelands (*ejidos*).

Over half of Mexico's 450-million-acre land base is composed of communally owned *ejidos*, a transformation that occurred after the Mexican Revolution of 1910–1920, when most of the large colonial plantations or estates were broken up and distributed among the nation's small farmer communities. Most of Mexico's 28,000 *ejidos*, however, are situated on semiarid and arid lands—deforested, eroded, and overgrazed, usually with no year-round water sources or irrigation. Overgrazing on these brittle lands has created a vicious cycle of degeneration—fewer plants, less forage, high runoff and low water retention, soil erosion, and fewer vital minerals, amplified by climate change.

The youth of the surrounding Jalpa Valley, with the exception of those who work at Vía Orgánica, seem to have very little interest in dryland farming, since in their experience farming amounts to struggling for bare survival. Many of the adults and youth in the area have already left, migrating to Texas, "*El Norte*" (usually illegally, since it's very difficult or impossible to get a US visa), in search of work and money to send home to their families. With millions of Mexican immigrants living and working in the United States illegally, it's risky to come back home and visit. Consequently, more and more migrants, once they cross the border, are becoming permanently separated from their families and rural communities.

In the Jalpa Valley, 170 miles north of Mexico City, where Vía Orgánica is located, almost every household is now dependent on multiple family members' low-paying jobs in the nearby tourist city (designated by the United Nations as a "World Heritage Site") of San Miguel de Allende. They get by on the substandard wages in the nearby foreign-owned *maquiladoras* or assembly plants, or else make ends meet with the remittances or money sent back home from their relatives working in the United States. In much of the state of Guanajuato, where San Miguel de Allende is located, and large parts of rural Mexico, the drug cartels take advantage of rural poverty in order to recruit unemployed youth into their ranks.

As I climb out of the truck with Humberto Fossi, our farm manager, a man named Dr. Juan Frias, a retired university professor and a noted expert on agaves and mesquite trees (as I learn later), comes up to us across the parking lot and introduces himself. Juan explains that he has just attended one of our weekly workshops at the ranch, this one dealing with organic compost techniques, and he wants to congratulate us all for the educational work that we're doing at Vía Orgánica.

Juan said he had been hoping to run into us because he wanted to discuss agaves. He pointed over to the dense grove of large salmiana agaves that we'd planted on a berm along the fence by the farm store and restaurant. Typically, our 8-foot-tall salmiana agaves, which require no irrigation, can weigh in at a ton or more at maturity (8 to 10 years old). They feature thorny, thick-skinned leaves, or *pencas*, the largest of which can weigh more than 40 pounds.

Like other agave farmers, or *pulqueros*, in our area, we only harvest the plant heart or *piña* (pineapple) and sap (honey water) of the mature agave to make a non-distilled alcoholic beverage called *pulque*. Before the arrival of European beer companies in the 1920s, *pulque* was the national drink in Mexico. Every village and urban neighborhood had a *pulqueria*. Our belief up until now had always been that the tough-skinned, spiny, and indigestible *pencas*, the leaves, were of no use except for compost. Farm animals typically won't eat the leaves of the agave, even if you chop them up with a machete, although they will eat the leaves or fronds of the nopal cactus if you scrape off (or burn off) the thorns. In the tequila and mescal industry, the discarded *pencas* and pulp (*bagasse*) of the agaves are either discarded and left to oxidize on the ground, or else burned.

"Ronnie," Juan Frias asked, "have you ever heard of pruning and finely chopping up the *pencas* of these *maguey* (agaves)? With a machine, you can shred them up finely, and then after fermenting the *pencas* anaerobically in closed containers, you can use the fermented agave silage as animal feed."

Rather surprised to hear this, I told Juan that no, I hadn't.

"There's a sheep and goat ranch," Juan went on, "not far from here, in San Luis de la Paz, called Hacienda Zamarripa, where the Flores González brothers have been planting and intercropping agaves and mesquite trees for over a decade, reforesting the parched *agostadero* (rangeland) and producing tons of fermented agave silage, which they are feeding to their animals as a primary forage. Their lambs actually survive and thrive solely on a diet of fermented agave and their mothers' milk, while 80 percent of the adult animals' diets are supplied by the fermented leaves of the agave."

Juan went on and explained that the production costs of this fermented agave are about 1 peso (5 cents) per kilo (2.2 pounds), making this the cheapest animal silage in Mexico, if not the world. The use of agave forage as a major part of the animals' diets, Frias explained, can eliminate overgrazing (the animals can graze for several hours a day instead of all day) and thereby help restore the rangelands during the dry season. Agave silage, 75 to 85 percent of which by weight is liquid or juice, keeps the goats, sheep, and lambs healthy and hydrated and increases their weight and value. Since the silage has high sugar content, the animal's meat has a distinct sweet taste. In addition, Juan explained, the entire system is organic, requiring no synthetic chemicals or pesticides.

The Zamarripa agave-mesquite agroforestry system (agaves and nitrogen-fixing trees), Juan added, planted in contoured or terraced rows, also reduces soil erosion, sequesters tons of carbon both above ground and below ground, stimulates pasture grass, provides habitat for birds and pollinators, and infiltrates and stores rainwater in the topsoil.

I was dumbfounded by what I was hearing.

Juan went on: "According to the US agave expert, Dr. Park Nobel, agaves can draw down and sequester more atmospheric carbon than almost any other plant or tree on Earth. And, of course, agaves can do all this with no irrigation whatsoever."

Still rather at a loss for words, I said, "No, I've never heard of fermenting the agave leaves, or an agave agroforestry system, but where can I find out more? Can we go and see this ranch?"

Juan assured us we could visit Hacienda Zamarripa. He promised to give his friend and fellow retired college professor, José Flores González, co-owner of the Zamarripa ranch, a call and get back to me. He explained that José Flores González was not only a rancher, with his two brothers, but also a retired mechanical engineer. José had designed and built the Zamarripa agave leaf shredder himself.

That night I checked out the Hacienda Zamarripa Facebook page, and became more and more excited. But searching on the internet I couldn't find anything about fermenting the indigestible leaves of the agave plant and turning them into animal feed. I did however find information about how agaves, a common desert succulent, along with the nopal cactus, another one of our plants on the ranch, ranked among the top 15 plants and trees in the world in terms of their ability to sequester large amounts of excess atmospheric carbon dioxide, in large part because of their unique system of photosynthesis called CAM. I googled CAM to find an explanation.

The Crassulacean acid metabolism (CAM) is a form of photosynthesis that evolved in desert plants as an adaptation to arid conditions. CAM allows a plant to carry out photosynthesis during the day, but only open up stomata on the underside of their leaves and exchange gases at night, thereby reducing the rate of evaporation. CO_2 is taken into the leaves of the agave and stored as Crassulacean acid. This enables the plant to store carbon and moisture and carry out photosynthesis even in prolonged extreme heat and drought conditions.

Carbon sequestration is a product of photosynthesis. Photosynthesis refers to plants' and trees' ability to breathe in or draw down carbon dioxide from the air or atmosphere and, using the energy of its solar energy collectors (the leaves), convert this CO_2 into carbon and oxygen. The plant, in this case the agave, releases oxygen back into the air, while using the carbon for its own growth. Besides building up its above-ground biomass, the plant or tree also transports and exudes a significant portion of its carbon in liquid form down through its roots system into the soil. Below ground this sugary liquid carbon feeds the universe of soil microorganisms that in turn make available the nutrients and minerals that the plant needs to grow. The ensuing increase in soil organic matter and humus sustains the life, biodiversity, and fertility of the soil and creates a spongelike environment around the plant that can infiltrate and retain rainwater. A 1 percent increase in soil organic matter means that the soil can retain an additional 16,640 gallons of water per acre.[1]

Since I was just finishing up the manuscript of a book on how to mitigate and reverse global warming with organic and regenerative food and farming, this encounter with Juan Frias, on a hot day in the desert, was extraordinary. In my research for *Grassroots Rising*, I hadn't come upon a real strategy for how to reforest and regenerate semiarid and arid desert lands without irrigation.

That night my imagination ran wild. As anyone who knows me can attest, I'm an incorrigible optimist. I like to think big. I love nothing better than to theorize, write, and give talks about systemic solutions to big problems, like the climate crisis and rural poverty. According to United Nations data, approximately 40 percent of the world's terrain is arid or semiarid, just like the land around Vía Orgánica and most of the land in Mexico and the Southwestern United States. Various varieties of agave are already growing, along with native nitrogen-fixing acacia trees such as mesquite, on billions of acres globally. In fact, there are already agaves growing on half of the arid and semiarid landscapes across the planet.

So, I projected, if you could grow enough plants, in this case billions of agaves and companion trees, grow them large enough, and interplant them on millions and millions of acres of the world's currently decarbonized and unproductive rangelands, you could conceivably draw down a critical mass of excess carbon from the atmosphere (where too much CO_2 contributes to climate change) and put it into the plants and trees aboveground, and into the soil belowground, where it belongs. By greening the desert and the drylands you could dramatically increase soil fertility, retain and store rainfall, restore landscapes and biodiversity, reforest semi-desert areas, regenerate rural livelihoods, and eventually restabilize the climate. I could hardly fall asleep.

The next morning, I got an email from Juan Frias saying we could visit Hacienda Zamarripa and the Flores González brothers in two weeks.

Early in the morning of July 11, four of us set off from Vía Orgánica to visit the Hacienda Zamarripa ranch located in the *municipio*, or county, of San Luis de la Paz, a two-hour drive away. Our group included myself; Humberto Fossi, our farm manager; Gerardo Ruiz, our permaculture design consultant and mesquite aficionado; and Dr. Juan Frias, our guide and Vía Orgánica's soon-to-be lead scientific advisor.

On the way to San Luis de la Paz, Juan filled us in on some of the history of the Zamarripa ranch. The Flores González family had owned and operated the ranch for generations, going back to the early nineteenth century. At one time the family ranch covered more than 1,000 hectares (2,500 acres), but over time, as the land was overgrazed and the native mesquite trees were cut down, Zamarripa became eroded and desertified, able to sustain fewer and fewer livestock. The family by necessity had to sell off parcels, little by little, until the ranch had shrunk down to 100 hectares (250 acres). The three Flores González brothers had no choice but to seek employment off the farm. José Flores González studied mechanical engineering, worked in a number of machine shops and factories, and became a university lecturer.

Over the years the situation in San Luis de la Paz, like most of the drylands in Mexico (which constitute 60 percent of the country), became more and more desperate. José and his brothers, Gilberto and Daniel, knew that the end was near for the family farm unless they could come up with a way to feed their sheep and goats, stop the loss of topsoil and fertility, and begin to restore the land before it degenerated into outright desert. Although sheep and goats (as well as other livestock and poultry) love the protein-rich bean pods, or *vainas*, of the mesquite tree, these tree pods are only available for a few months of the year, and the quantity of *vainas* are limited, since most of the native mesquites in San Luis have been cut down over the years. If animals are starving, they will eat a bit of non-fermented agave leaves chopped up with a machete, but these leaves contain a chemical compound called saponin and are basically indigestible.

Flores González speculated that perhaps pruning and chopping up the leaves of the agave plant, and then anaerobically fermenting the leaves in a closed container (transforming the saponins into carbohydrates and sugar), could sustain the family's livestock through the dry season. Meanwhile intercropping and reforesting his lands with mesquite seedlings as a companion tree could fix the nitrogen and nutrients in the soil that the fast-growing agave and its shoots, or *hijuelos*, need to grow and multiply. A single mother plant of agave can produce up to 40 *hijuelos* during its lifetime. These *hijuelos* can be readily removed from the mother plant and transplanted, producing a new agave.

Taking advantage of his academic training, José designed a shredding machine that could be hooked up to a tractor to give its cutting blades enough torque and velocity to slice and chop the tough leaves and spiny thorns of the *pencas* into small enough pieces that once fermented the animals would consume.

Along the bumpy, dusty road leading to Zamarripa, the landscape was desolate. Much of the pasture and rangelands in the Jalpa Valley, where Vía Orgánica is located, are in pretty bad shape, but this was worse. It looked like the moon, with gullies and erosion visible on the slopes. There were very few trees or bushes in sight and the few sheep and goats we passed were skinny and ragged.

As soon as we passed through the gates at Zamarripa, things looked different. Still an arid landscape this time of year, but there was more mesquite, mainly younger trees, and lots of salmiana and americana agaves, planted in contoured rows, growing alongside the agaves and nopal cactus. There were even some small patches of grass under the shade of the mesquites and agaves.

We got out of the truck to take some photos. Gerardo, our permaculture expert, said that the contoured rows and reforestation design looked pretty good. "But," he said, "I don't understand why the agaves and the mesquite are planted so close together."

A little further down the road we got our answer. The mesquites here were older and taller than in the first area we had seen, and there were lots more agaves, 2,000 per hectare according to Juan, and amazingly the agaves were growing up right next to the mesquites. The shade of the mesquite trees, which would stunt the growth of certain agaves (such as the blue agaves for tequila and the sisal/henequen agaves for fiber and textiles), seemed to have no impact on the more rustic, hardy varieties of agave (*salmiana* and *americana*) that were growing here.

These Zamarripa agaves were the same native varieties we used for *pulque* at Vía Orgánica. Juan explained that these varieties grew throughout much of Mexico. This meant, Juan went on, that you could have an agroforestry system like Zamarripa's (2,000 agaves and 400 nitrogen-fixing trees for each 2.5 acres) that could produce massive amounts of silage but still maintain a sustainable balance between nitrogen and nutrient levels in the soil and the fast-growing high biomass agaves. And, he added, we were looking at an agroforestry system that was drawing down and sequestering large amounts of carbon dioxide from the atmosphere, even in poor soils with no irrigation other than the seasonal rainfall of 10 to 15 inches per year during the brief rainy season. We learned later that 2,000 agaves and 400 nitrogen-fixing trees in a 2.5-acre area could sequester 140 tons or more of atmospheric CO_2 over a 10-year period.

We drove along further and stopped at the barn where two of the Flores González brothers (José and Gilberto) came out to greet us. The tallest of the brothers, Gilberto was dressed up in a khaki-colored uniform that looked like something out of a Pancho Villa / Mexican Revolution movie. Except that Gilberto was skinny, not portly like Pancho Villa. José with glasses and a baseball cap looked rather more contemporary.

José and Gilberto shared with us the history of the ranch. They had inherited Zamarripa from their father. What had once been a relatively sustainable ranch for their grandparents had become degraded, like most of the desertified drylands of Mexico. The brothers could see they were losing topsoil every year and needed a native plant that could hold the soil in place. They realized that the best plant to stop the erosion and potentially feed their livestock would be the hardy agave. But their early attempts to chop up the agave leaves by hand with machetes and ferment them in plastic garbage bags met with mixed results.

They learned, they told us, that the smaller, shredded agave leaf pieces fermented much better than the larger ones. They saw clearly that sheep and goats would only eat the smaller, finely chopped-up fermented agave pieces. They also learned, by trial and error, that placing a vacuum cleaner and sucking out all the air at the opening of the garbage bag before sealing the bag improved the anaerobic fermentation. They realized that manual chopping was too labor intensive. A group of three or four workers with machetes could only chop 750 pounds of agave leaves per day, and most of the pieces were not small enough. They concluded they needed a machine or mechanical chopper to get smaller, uniform pieces, and a greater volume of silage. They tried out various farm choppers and grinders that were used for corn and other grains, but nothing worked really well with the tough agave leaves.

So the Flores González brothers sat down in their workshop and designed and built the Zamarripa shredder. It took them almost a year to perfect their design, but in the end it worked well. Hooked up to a tractor to power the blade at 1,200 RPMs, the shredder could chop up over a ton of agave *pencas* in an hour. Once fermented, a ton of agave silage can sustain a herd of 50 adult sheep for 10 days during the dry season. Twenty-four tons of silage, easily pruned from 2,000 agaves on a single hectare (2.5 acres) every year, would sustain these 50 sheep over the entire year.

The Flores González brothers walked us over to the open-air agave processing area next to the barn. The Zamarripa shredder was already hooked up to the driveshaft of an old John Deere. What looked like a pickup truck load of pruned agave *pencas* was lying on the ground next to the shredder. José signaled for the tractor operator to begin and the engine roared to life. For a minute or so the driver warmed up the engine and gradually opened the throttle until the old tractor was running at top speed. José told two men in gloves and masks who were to feed the agave into the machine to begin, and with a loud noise, sort of like a chain saw, the giant agave *pencas* were shoved into the blade of the shredder, which began spitting out a steady stream of what looked like green coleslaw. The wet mound of coleslaw in the metal container under the shredder grew higher and higher until José yelled over the noise for the tractor operator to cut the engine.

Next, the workers started shoveling the shredded agave leaves into large black garbage bags. After they filled a bag they would attach a vacuum cleaner nozzle into the end of the bag, suck out the excess oxygen, and then twist and close up the bag with a plastic tie. They then carried the bags of shredded agave into the *bodega* or storage shed, where hundreds of bags were either fermenting or already fermented, ready for feeding to the livestock. The workers carried out several bags of fermented agave and opened them up for us to see. After 30 days of fermentation the green coleslaw had turned into a golden-colored silage. The indigestible saponins in the leaves had now been turned into sugar and carbohydrates. José reached into one of the bags and brought the *foraje* mixture up to his nose. We all did the same. The rich smell reminded me of sauerkraut or kombucha.

Our next stop was the fenced-in feeding area for the adult sheep and goats who had been grazing all morning in the outlying, rather sparsely vegetated, end-of-dry-season rangeland. The animals were all waiting eagerly at the gate for their midday meal. Two workers with wheelbarrows filled with bags of silage began puncturing the plastic garbage bags and shoveling the fermented silage into the slotted feeding troughs. I couldn't help thinking to myself that they probably weren't recycling these plastic garbage bags at Zamarripa and that we'd have to come up with a different system back at Vía Orgánica.

Once the long feeding troughs, with room for 25 animals on each side, were filled with silage, one of the workers opened the gate that connected the feeding pad lot to the grazing areas. The sheep and goats literally poured in, rushing over to the agave feeding troughs. Even the two border collie sheep dogs joined in on the action. There was no doubt that the animals loved the sweet fermented agave. They were gobbling it down like candy. Several of the goats climbed up on all fours into the troughs.

I realized that the Flores González brothers had achieved an amazing breakthrough. They had discovered, as no one had before, how to ferment and transform an enormous amount (potentially billions of tons) of previously unused and discarded agave plant biomass into a valuable feed supplement. Mexico has several billion agave plants growing, either cultivated, semi-cultivated, or growing wild, with the potential for many billions more. At least half of the biomass of these large plants are their leaves or *pencas*. The heretofore discarded *pencas* of these agaves pruned on a yearly basis could produce literally billions of kilos of fermented animal feed every year, enough to sustain all of the livestock in Mexico during the long dry season, while supplementing their diet during the rainy season when there's more grass and forage.

At lunch we dined on delicious grilled lamb tacos, the sweetest and most tender I had ever tasted. I told José Flores González and his brother Gilberto that they deserved the Nobel Prize for agriculture for what they had discovered. I promised we would deploy the agave mesquite agroforestry system back at Vía Orgánica and spread the news of "agave power" far and wide.

Before we go any further into this incredible tale of agave power, perhaps a little background information is in order. How did my wife, Rose, and I, lifetime activists and directors of the Organic Consumers Association (OCA), end up in the semi-desert highlands of north-central Mexico? OCA's national office and Agroecology Center are located in the far north woods of Finland, Minnesota, where we have seven months of snow and frigid weather every year. What motivated us to recruit a group of Mexican and international activists to set up a farm school called Vía Orgánica, the "Organic Way?" And what motivated us and our team to spend over a decade building an organic farm school, restaurant, grocery store, and permaculture center in a poverty-stricken high-desert community with no wells or irrigation, extreme temperature variations, and an eight-to-nine-month dry season with no rain whatsoever?

The short answer is that Rose and I, like many foreigners who have become residents, love Mexico and Mexican culture, especially the hardworking farmers and rural villagers who live in places like Membrillo, where we now live eight months a year on the Vía Orgánica Ranch and Agroecology Center.

We're also politically motivated, in my case as an activist, organizer, fundraiser, and writer ever since the Vietnam War and the Civil Rights Movement of the 1960s. A driving motivation in my life has always been to work and stand in solidarity with the exploited and marginalized, whether in the United States or abroad. Rose similarly has been active as an antiwar, organic food, and solidarity activist for more than 30 years. My first trip to San Miguel de Allende, a picturesque, cobblestoned former Spanish colonial city (and UN World Heritage Site) 15 minutes from the current location of the Vía Orgánica Ranch, was in 1967, when I hitchhiked all the way from my student/hippie commune in Houston, Texas. I fell in love with San Miguel and Mexico at first sight, especially the people and the culture, which reminded me of my Cajun grandparents and our traditional family farm in East Texas, where I was fortunate enough to spent the weekends and summers of my youth. Later, in the 1980s, Rose and I fell in love with all the countries in Central America, especially Guatemala.

Before founding the Washington, DC–based Pure Food Campaign (1992–1998) and the Organic Consumers Association (OCA; 1998–present), Rose and I spent five incredible years in Guatemala and in the conflict zones of Central America—Nicaragua, El Salvador, and Honduras. During this time, I was working as a freelance war correspondent and activist, focused on peace, social justice, human rights, refugee support, and solidarity issues, while Rose was a photographer, nonfiction children's book author, and solidarity activist as well. Rose and I coauthored five children's books on Guatemala, El Salvador, Costa Rica, Nicaragua, and Cuba in the late 1980s. We also spent a year in Costa Rica, managing a rustic eco-lodge, working on a popular travel book, and doing ecotourism research.

I truly love my work in the United States, writing, editing, and fundraising for the OCA, and campaigning on food, climate, anti-GMO, natural health, and antiwar issues. But my overriding passion nowadays is working with the small farmers, especially youth, and rural communities south of the border. These are the people who are forced to choose between leaving their families and communities and crossing the border to seek work in the United States to earn money to send back to their families, or to stay home and literally struggle for economic survival.

As Rose and I will tell you, the solution to the "immigration crisis" in the United States (and Europe) is not to build an impenetrable wall, or, on the other hand, to have "open borders" and encourage people to leave their families and rural communities, but rather to work and organize so as to create rural prosperity and peace back in migrants' home communities, utilizing organic and regenerative food and farming.

At our Vía Orgánica farm, restaurant, and store we employ 53 Mexican nationals. Our staff are predominately young women and men, recently graduated agronomists or local farmworkers, builders, chefs, teachers, and waitstaff, nearly all of whom come from low-income backgrounds. Because we pay relatively well, respect our team members, and offer youth, including young women, the experience they need to become future organic leaders, none of our staff are interested in leaving Vía Orgánica and crossing the border illegally.

What do people in Mexico and on the farm think of us? They all love Rose, my red-haired, 100-percent Irish life partner, homeopathic healer, chef extraordinaire, and OCA general manager. Rose no doubt reminds many of them of their mother. But what do they think of me, the old gringo, the grey-haired ponytailed ex-hippie, now director of Vía Orgánica and the Billion Agave Project?

In meetings across rural Mexico, or in gatherings in Mexico City with activists who work with our Vía Orgánica and *Asociacion de Consumidores Organicos* network in coalitions like *Millones Contra Monsanto* (Millions Against Monsanto), *Sin Maiz No Hay Pais* (Without Corn There is No Country), or the network of *Tianguis Organicos* (organic farmers markets), it's customary to start off gatherings by going around the room and introducing yourself. I'm often the only gringo or foreigner in these gatherings.

When I introduce myself, I say, "I'm Ronnie Cummins. I'm originally a *Tejano*, born and raised in Texas. But since my ancestors robbed Texas from your ancestors, I'm actually a *Mexicano*." My intro never fails to generate smiles and solidarity.

Although I've often been criticized as a "utopian" thinker and incorrigible radical optimist, I must admit that I am not immune to periodic bouts of discouragement. In the early summer of 2019 in Mexico, at the end of the long dry season, just as I was finishing up my manuscript *Grassroots Rising: A Call to Action on Food, Farming, and Climate* and preparing to send it in to my publisher, I began to doubt whether my book's central message would really resonate with readers and activists. Even though it was true—that organic and regenerative food, farming, and land use (in combination with renewable energy and conservation) can fix the climate and rural poverty—most climate activists were still discussing the climate crisis in isolation from food and farming, while most organic farmers and consumers were not connecting the dots between healthy food and environment and the climate. Even fewer seemed to believe that we could actually reverse global warming, instead of just slowing it down.

While I was writing the manuscript for *Grassroots Rising* in 2018–2019, our US-based organization, the Organic Consumers Association, and our international network, Regeneration International, were working diligently with food-savvy youth in the Sunrise Movement in the United States to recruit US farmers into a new national populist coalition called Farmers & Ranchers for a Green New Deal. We created, at least temporarily, North and South, an amazing alliance of rural conservatives, family farmers, ranchers, and libertarians, united together with organic farmers, many of whom were transplanted city folks and considered themselves environmentalists or political progressives. Generating excitement for the coalition, Vermont Senator Bernie Sanders, a leading candidate for the Democratic Party nomination for president at the time, and a lifelong populist like myself, had actually begun talking about the need for a new food and farming system in the United States that was organic and regenerative. Bernie's platform on agriculture, in fact, was largely written by several organic farmers from the Midwest who were supporters of our Farmers & Ranchers for a Green New Deal. Unfortunately, although I guess predictably, Bernie was pushed aside in early 2020 by the Democratic Party establishment. The Green New Deal and the notion of a populist and organic revolution in US food and farming, at least for the foreseeable future, died a premature death.

I like to think of myself not just as a radical populist, but also an internationalist. Throughout my entire life as an activist, going back to the 1960s, I have always tried to prioritize both US and international work. A few years after the formation of OCA, in 2002, as soon as we could raise enough money to do so, we set up our sister organization in Mexico, called "Vía Orgánica." At first our small staff was based in the highlands of San Cristobal de las Casas, Chiapas, near the

Guatemala border, scene of the Zapatista Indigenous uprising of 1994. In Chiapas Vía Orgánica focused on anti-GMO corn issues, organic women's community gardens, bio-piracy, and seed saving with Indigenous Zapatista communities. Finding it extremely difficult to promote organics and set up a food hub in the middle of a conflict zone, where 60,000 Mexican soldiers were stationed to keep the Zapatistas in line, we moved our operations in 2005 to Mexico City and San Miguel de Allende, Guanajuato, a high-desert municipality in north-central Mexico, 170 miles north of Mexico City.

Over the years in San Miguel and Mexico City, Vía Orgánica became one the leaders in the anti-GMO and pro-organic movement in Mexico. Our staff of Mexican citizens and international volunteers grew to 70 people, with an organic restaurant, grocery store, website, newsletter, and farm school gaining recognition and support from farmers and consumers alike. But until we discovered "agave power," our results, in terms of spawning an organic agricultural revolution, no matter how well intended, were limited.

It is very difficult nowadays for a small farmer or rancher in Mexico to farm traditionally, or to make the transition to certified organic production, and earn a decent living. Drylands subsistence farming is possible in theory, but markets are limited, irrigation wells are scarce, and the cost of living in Mexico, even though much lower than the United States, is constantly rising. Until our epiphany at Zamarripa, I was beginning to doubt whether we could really make a difference in a poor country suffering from climate change, degraded landscapes, and an ever-rising cost of living. But now, I realized, perhaps agave power could change all that, and not just in Mexico, but in most of the arid and semiarid regions of the world.

CHAPTER TWO

Climate Science, Skepticism, and "Solutions"

In early July, 2019, Ronnie called me and told me about the agave agroforestry system based on turning previously inedible parts of the agave and mesquite into high-quality animal feed by fermenting them. He asked for my opinion on scaling it up.

Ronnie and I regularly talked, bouncing ideas off each other. This is how we started the steering committee that became Regeneration International in 2015. Ronnie, Vandana Shiva, Steve Rye, and I were giving presentations about organic agriculture and climate change at an event at the Rodale office in New York in 2014. Afterward, we proposed having our organizations partner to build a more significant global movement.

The initial discussions were between Ronnie, Mercola CEO Steve Rye, Dr. Vandana Shiva and Dr. Hans Herren from the Millennium Institute, and myself, as president of IFOAM - Organics International. We devised the plan to form Regeneration International and launch the global regeneration movement. Hardly anyone had heard of regenerative agriculture before then. It is in the news every day worldwide now because of our initial actions.

Regeneration International is the largest and most significant regenerative organization on the planet, with over 570 partners in 75 countries in Africa, Asia, Latin America, Australasia, the Pacific, North America, and Europe. We are growing every month and continue to lead this global movement.

Ronnie and I would regularly call each other to discuss significant issues and developments. When he told me about his visit to the Flores González brothers' farm and Dr. Juan Frias's information, I knew he had found a vital agroforestry system.

I immediately realized this was a significant breakthrough in designing new regenerative systems for most climates, not just arid and semiarid areas. It meant many species of native plants could be fermented to remove saponins, lectins, and other toxic compounds and could now be used as high-quality feed. It would allow the planting of native species to regenerate ecosystems that could be selectively harvested for feed.

I am a great believer in ground-truthing claims that are made about agricultural systems. I have visited thousands of farms and ranches over the last 50 years in more than 100 countries on the six arable continents and seen them firsthand to verify their claims.

I need to see them to ensure that the reported benefits, yields, and improvements are real. There are numerous claims made in agriculture about the benefits of various systems. Many of them are exaggerations or outright lies, especially in industrial-agricultural systems. These systems create enormous health, environmental, and social problems without evidence of improvements over nature-based regenerative systems.

Due to COVID-19 restrictions, I could not leave Australia without a special government permit, which was rarely granted. It was even harder to return with a waiting list of over 30,000 Australian citizens wanting to return to their country.

Finally, in December 2021, the government ended the restrictions, and we could leave Australia without a permit. Julia and I caught one of the first flights to the United States. From there, we headed to Mexico to the World Heritage city of San Miguel de Allende near the Vía Orgánica Ranch.

San Miguel de Allende is one of the oldest European cities in the Americas. It was built at the start of the Spanish colonial invasion to mine silver, which was exported from Acapulco and sold to the Ming Dynasty in China via Manila in the Philippines. This amassed great wealth for the Spanish ruling classes. They stole all the lands and enslaved the Indigenous peoples to work in brutal conditions in the mines or in the fields. The trees were cut down to smelt the silver, drying out the region and changing its climate and ecosystems.

Despite its cruel history, San Miguel de Allende is a stunningly beautiful historical city that was not destroyed by developers building the ugly modern concrete monstrosities that have replaced much of Mexico's vernacular. The inner city has miles and miles of well-preserved traditional houses, churches, and other historic buildings and is registered as a UNESCO World Heritage Site.

It is a city that most people instantly love for its charm and historic streetscapes. After two years of not being allowed to travel, it was wonderful to be back in this special place, enjoying good food and great company. Julia and I always enjoy staying at Casa Angelitos, run by our friends Roger Jones and Rosana Álvarez. This boutique hotel with stunning views over the city is like our second home in Mexico.

Ronnie and his wife, Rose, fell in love with San Miguel de Allende and moved there with their son, Adrian. They lived there for many years and later spent part of the year in Minnesota and part in Mexico.

They started several organic projects there. One of the most important is the Vía Orgánica Ranch, just a few miles from the city.

Ronnie and Rose had been very active in the two years since I'd last visited the ranch. They moved a restaurant and shop they had in town to the ranch. While eating fresh organic food from the ranch over lunch with old friends such as Ercilia Sahores, our Latin America coordinator, and Humberto Fossi, the ranch manager, I met Dr. Juan Frias for the first time.

Ronnie had started trialing the agave agroforestry system by planting many acres with agaves and mesquite trees. Ronnie and Juan showed me the planted fields. I was shown the machines used to chop up and ferment the agave and mesquite leaves to make the silage. It was the dry season, and the livestock was penned up for the evening and being fed the silage. They relished it. I could see from the bloom on their coats that they were in excellent condition. It was evident to me that this process made formerly inedible plants very edible and nutritious.

Most ecosystems have periods of low pasture growth in cool and dry seasons when they are overgrazed, causing environmental degeneration. The fermented forage, as silage, can be used as feed during these periods to take animals off the pastures and native ecosystems to allow them time to recover. Recovery time is essential to successful regenerative grazing systems. This is a time when livestock lose condition due to lack of feed and suffer. The agave and mesquite silage benefited the environment and the livestock.

This system filled a significant gap in the evolving area of agroforestry and silvopastoral systems.

Agave agroforestry can readily be scaled up on hundreds of millions of acres of underutilized or nonproductive landscapes, transforming currently degraded, deforested, overgrazed, and poverty-ridden lands. These are the lands that will never be able to sustain adequate food production, livestock, and animal husbandry, or adequate incomes for the billions of people who live there, unless the farmers and ranchers can be motivated (and paid) to restore their lands and deploy this game-changing agroforestry system.

Very significantly, it can be scaled up to all continents using their endemic legume trees. There are thousands of species of legume trees. Every continent—Africa, Australia, Eurasia, North America, and South America—has many hundreds of species adapted to their arid, semiarid, and higher rainfall regions.

Deep-rooted legume trees such as mesquite, acacia, sesbania, inga, leucaena, and faidherbia offer an effective way of drought-proofing dry lands and dry seasons as their roots extend deeper into the soil than grasses to find water.

Agaves require no irrigation. They efficiently store seasonal rainfall and moisture from the air in their thick leaves and stem or heart. This enables the plant to grow and produce significant biomass even under prolonged droughts. Various species of agave are now naturalized on all continents, so scaling up this agroforestry system with native legume trees and agaves will not cause an environmental weed problem. The agroforestry system is a way to manage them effectively.

Cover crops and perennial pastures can be planted to produce more fodder and increase soil fertility.

The United Nations Convention to Combat Desertification states that 500 million pastoralists herd their animals across highly diverse, grass-dominated rangelands that cover one-third of the Earth's land surface.[1]

Most of these rangelands are unsuitable for annual crops as tillage erodes and damages the soil. In many cases, the amount and timing of rainfall are inadequate to grow row crops such as grains, fruits, and vegetables without irrigation. They are only suitable for grazing livestock, the residents' primary livelihood. Many of these ecosystems have been badly degraded by incorrect grazing, resulting in eroded and human-created desert landscapes, with the animals kept in poor condition due to a lack of feed in the drier periods.

By implementing these agroforestry practices, farmers and ranchers can begin to regenerate arid landscapes and improve the health and productivity of their livestock. The advantage of the agave agroforestry system is that apart from watering the seedlings at the initial planting, this system does not need irrigation. After several years, the agroforestry system can provide affordable food and income for families and improve their livelihoods.

It also delivers valuable ecosystem functions such as reducing soil erosion, recharging water tables, and drawing down and storing large amounts of atmospheric carbon dioxide in plant biomass and soils, both above and below the ground.

These regions have some of the highest food insecurity and malnourishment levels, especially among children. They have regular droughts that turn people into refugees, migrating to other regions for food and employment to survive. The increase in extreme weather events caused by climate change worsens this.

The agave and other agroforestry systems can end hunger, poverty, conflict, and environmental degradation by regenerating endemic native plant and animal biodiversity to provide food and income for local residents. Scaling them up is also an effective strategy for mitigating and adapting to climate change.

Climate Change

Climate change is a significant issue for all agricultural systems because, among other reasons, it causes increases in extreme weather events. The climate is influenced by the amount of solar energy that enters or leaves the Earth's soil, atmosphere, and oceans. This energy causes temperatures to rise or fall, produces our normal seasons, and impacts extreme events like storms, floods, and droughts.

The oceans and the atmosphere are more than 2.2°F (1.2°C) warmer than in 1750, the start of the Industrial Revolution. This is due to an energy imbalance where more solar energy is retained as heat than is emitted back into space. When the same amount of energy received from the sun as electromagnetic (solar) energy in the visible light and infrared range is emitted back into space, the temperature is stable, and the climate is in equilibrium. The planet heats up when more energy is retained and reflected around the atmosphere and oceans. When more energy escapes the planet than is received, the planet cools and enters into an ice age.

Climate Forcings

Events that change the solar energy equilibrium are called climate forcings. Changes in the sun's brightness, long-term sun cycles, minor variations in the shape of Earth's orbit over thousands of years, volcanic eruptions that inject light-reflecting particles into the stratosphere, methane from melting permafrost, and forest fires from lightning strikes are natural climate forcings. People cause climate forcings through numerous activities. The rising concentration of atmospheric carbon dioxide and other greenhouse gases through the burning of fossil fuels, deforestation, loss of soil organic matter, deliberately lit forest fires, methane leakage from gas wells and animal factory farming, and nitrous oxides from synthetic fertilizers trap heat and energy, warming the planet. Dust haze, particulate pollution, and other human-made aerosols block incoming solar radiation, reducing the incoming energy and therefore cooling the planet.

Scientists are debating various theories and models of why this happened in the past. The combination of the long-term variation of the Earth's slightly eccentric orbit called the Milankovitch cycles and the long-term Bray and Eddy sun cycles offer plausible explanations for past climate change events.

The Bray cycle is about 2,450 years from beginning to end. The Eddy cycle is 980 years long. The Milankovitch cycles are around 100,000 years long. While not perfect, the combination of these three cycles into a model correlates well with 100,000-year cycles of ice ages and warm periods. The model correlates with minor climate events such as the Roman Warm Period, around 100 CE; the Dark Ages Cold Period, around 650 CE; the Medieval Warm Period, around 1100 CE; and the Little Ice Age, around 1650 CE. The supporters of this theory state that the current warming trend is just part of these cycles.

One of the central tenets of science is that correlation is not causation. Just because two things correlate does not mean one causes the other, as it could be a coincidence. Research has shown that the minor warm and cool cycles, such as the Medieval Warm Period and the Little Ice Age, were restricted to the North Atlantic regions, and these temperature fluctuations did not occur in equatorial or Southern Hemisphere regions. Other theories have been put forward for these local temperature fluctuations, such as changes in ocean currents, Icelandic volcanic eruptions, and the Amazon rainforest regeneration after diseases introduced by European colonists decimated the Indigenous civilizations. The reality is that the causes of past climate events are strongly contested by scientists supporting multiple theories, and there are substantial disagreements between the supporters of the theories.

There has been no discernible long-term increase in solar radiation since precise monitoring began in the 1970s, which casts doubt on these cycles being the cause of the current global temperature increase.

Anthropogenic Greenhouse Gases

The dominant theory for the current cause of climate change is the increase of anthropogenic (man-made) greenhouse gases, particularly carbon dioxide. Man-made greenhouse gases include carbon dioxide (CO_2), methane (CH_4), nitrous oxide (N_2O), trichlorofluoromethane (CFC-11), dichlorodifluoromethane (CFC-12), carbon tetrachloride (CCl_4), carbon monoxide (CO), nitric acid (HNO_3), and tropospheric ozone. They absorb radiant heat and energy and reflect it in the atmosphere, causing an amplifying effect on water vapor, the main greenhouse gas. This is radiative forcing, and despite skeptics saying there is no evidence that CO_2 causes warming, this forcing has been detected and measured since the 1970s.

Water vapor is responsible for 75 percent of the greenhouse gas effect; however, it does not persist, and the excess heat quickly goes out into space. Carbon dioxide is responsible for 20 percent of the greenhouse gas effect and is highly persistent, lasting over a thousand years. The other gases—methane, nitrous oxide, and halocarbons—account for 5 percent of the greenhouse effect and are not as long-lasting. The models show that the heat-amplifying effect of CO_2 stabilizes and amplifies water vapor as the main greenhouse gas so that heat does not readily escape into space. This leads to a net increase in energy and heat fueling our weather systems.[2]

Ice cores from drilling in Greenland and Antarctica have been analyzed to determine the temperatures, carbon dioxide, and methane levels over the last 800,000 years. CO_2 and methane track the rise and fall of temperature with the ice age cycles. However, in most cases, the rise in CO_2 occurs hundreds to several thousand years after the temperature rises, showing that it is the result of rising temperatures, not the cause. Climate change skeptics use this to state that CO_2 is not the cause of temperature rises.

For the first time in 800,000 years, due to the burning of fossil fuels, the loss of soil organic matter, and the clearing of forests, the level of CO_2 is rising faster than the temperature. It has risen much faster in the last hundred years than ever in the past 800,000 years.

Scientists have researched how CO_2 and other greenhouse gases drive temperature increases. NASA launched the IRIS satellite in 1970 to measure infrared radiation. Infrared is part of the spectrum of solar radiation that makes heat. The Japanese Space Agency launched the IMG satellite in 1996, which recorded similar observations. The Japanese data over the past 26 years found a decrease in radiation going out into space at the infrared wavelength bands at which greenhouse gases such as carbon dioxide (CO_2) and methane (CH_4) absorb energy. The measurements were direct evidence that proved the increase in heat and energy absorbed and radiated by these gases.

These results have been confirmed by subsequent research using more recent satellite data, showing that anthropogenic greenhouse gases trap energy and heat and are a significant cause of climate change. This has added an extra 4.1 W/m^2 (watts per square meter) of energy into the atmosphere since 1750, the start of the Industrial Revolution.

Problematic Climate Models

It is complicated to translate this information into how much and when it will heat the planet and the resulting increase in extreme weather events such as hurricanes, floods, and droughts. Scientists have been producing models using multiple methodologies to make climate change and temperature predictions based on these measurements. The problem for scientists is that the complexities of climate mean that models use different assumptions and combinations of factors such as incoming energy, absorption of energy, and refracted energy distributed through the atmosphere and albedo, where the energy is reflected out to space. The changing atmospheric dynamics and weather make the simulations extremely complex with variable results. The models also have significant uncertainties due to the difficulty of accurately predicting the climate feedback mechanisms associated with the interaction of oceans, vegetation, clouds, and water vapor with the greenhouse effect.

Using models to make predictions as the basis of national and international climate change policies is highly problematic. Unfortunately, multiple unreliable predictions based on previous models and promoted by celebrities and scientists—such as substantial temperature increases that would cause all glaciers to melt away, an ice-free Arctic, permanent droughts, and massive sea level rise, some of which were supposed to occur decades ago—have proven to be gross overestimations. Climate change skeptics have promoted these to discredit the whole concept of climate change as a "climate scam." The skeptics continuously raise these failed predictions and use them to call the current predictions alarmist.

There are significant problems with what the models leave out, such as the profound effects that soil organic matter and vegetation have on hydrology, evaporation, transpiration, and local and transcontinental weather. The omission of these significant climate forcings is one of the reasons why the current models have failed in terms of overestimating the amount of global warming and underestimating the damaging local climate extremes such as droughts, hurricanes, and floods.

Ocean Forcings

According to the National Oceanic and Atmospheric Administration (NOAA), more than 90 percent of the warming on Earth over the past 50 years has occurred in the oceans. The oceans and the atmosphere are already more than 2.2°F (1.2°C) warmer than at the start of the Industrial Revolution. The ocean's heat energy is a significant driver of weather.[3] Even if CO_2 levels went down to the 1750s level of 280 parts per million (ppm), it would take many decades for the heat in the oceans to dissipate. The oceans store 91 percent of the excess heat trapped by greenhouse gases and 31 percent of human emissions of CO_2. These greenhouse gases and heat have been taken deep underwater by currents that are releasing this heat and some CO_2 to the surface.

Researchers have found that the rate of the ocean heating up has increased significantly since the 1990s. Most of the ocean warming and thickening is in the subtropical Atlantic and Pacific Oceans and the Southern Ocean. These systems are poorly understood, so modeling future temperature increases and climate scenarios is fraught with uncertainties.[4]

What is certain is that the extra ocean heat is fueling stronger storms, especially hurricanes (tropical cyclones). Research into hurricanes from 1979 to 2017 found an 8 percent increase in intensity of category 3 to 5 storms every decade.[5] According to NOAA's Geophysical Fluid Dynamics Laboratory, the extra warming means that there will be an increase in very intense category 4 and 5 hurricanes.

As I write this in October 2023, Hurricane Otis has broken all records, going from a tropical low to a category 5 storm in a few hours after passing over a body of warm water, devastating Acapulco. This had never happened before, leaving the meteorologists completely unprepared so that no warnings were given for residents to evacuate or take adequate shelter.

A few days before Otis, on the other side of the Pacific in the Southern Hemisphere, Cyclone Lola formed in warm waters off the Solomon Islands. It is rare for a cyclone to form in October before the official start of the South Pacific cyclone season in November. Lola quickly became a category 5 before hitting and devastating northern Vanuatu. It was the first time a category 5 cyclone had been recorded in the South Pacific in October.

The warming of the atmosphere increases its ability to hold more water vapor. Studies suggest that hurricane precipitation increases by 10 to 15 percent for a 3.6°F (2°C) global warming scenario.[6] Similarly, there is an increase in storms creating destructive floods across all regions of the planet.

Deforestation

Man-made greenhouse gases are just one type of climate forcing. Forests, vegetation cover, and soils profoundly affect local and global temperatures, the transpiration of water vapor, and atmospheric and terrestrial hydrology. Forests absorb heat and energy through photosynthesis, shade the ground, and cool the planet through the transpiration of water vapor. Deforestation results in the land absorbing sunlight and heating up, making this a significant climate forcing.

According to Our World in Data, 1.5 billion hectares (3.7 billion acres) of forest have been cleared over the last 300 years—since the beginning of the Industrial Revolution. That's an area 1.5 times the size of the United States. Ten million hectares (25 million acres) of forest are destroyed every year. Ninety-five percent of this deforestation occurs in the tropics.[7]

The European Investment Bank estimates that around 15 to 18 million hectares (up to 45 million acres) of forest are destroyed yearly, with 2,400 trees cut down each minute.[8]

This loss of forest cover is a significant climate forcing and contributor to increasing global temperatures.

A high proportion of this deforestation is driven by consumers in the world's richest countries: Cleared land is used to produce GMO soy and maize to feed cruel confined animal feeding operations (factory farms) in western Europe and East Asia. These feeding systems are incredibly inefficient. They need 10 tons of vegetable protein to produce 1 ton of animal protein. There is no justification for clearing these forests to "feed the world" when they are not feeding the food-insecure. This industrial agriculture system grossly wastes land and resources to provide luxury commodities for the world's wealthiest consumers. The same applies to the beef, vegetable oils, cocoa, coffee, and paper produced on deforested land and exported to the Global North. They are not produced to feed the undernourished; they are luxury commodities for the world's richest consumers.[9]

The IPCC (Intergovernmental Panel on Climate Change) states that aboveground plant growth is now beginning to increase globally due to increases in temperate and boreal forests and despite tropical forests declining due to deforestation.[10]

Researchers in the United States and China used satellite data to analyze the biophysical effects of forests on local climates. Their results showed that tropical forests have a strong cooling effect throughout the year; temperate forests show moderate cooling in summer and moderate warming in winter with net cooling annually; and boreal forests have strong warming in winter and moderate cooling in summer with net warming annually.[11]

The increasing forest types are boreal forests, which have a net warming effect, and temperate forests, which have moderate net cooling, although the recent devastating fires in boreal forests of Canada, Alaska, and Russia, and temperate forests in Mediterranean climates, may have significantly reversed this. Ninety-five percent of deforestation occurs in tropical forest ecosystems. These forests contribute to a strong cooling effect throughout the year. These highly biodiverse tropical ecosystems are being cleared at unsustainable rates for industrial commodity agriculture, such as GMO maize and soy, palm oil, and beef production. Clearing these forests means their cooling effect is replaced by heat islands, contributing to global warming.

Numerous studies have shown that tropical deforestation increases local daytime surface temperatures by 1.8 to 3.6°F (1 to 2°C) or more. Deforested areas in the Amazon were found to have warmed by as much as 5.4°F (3°C) locally, most significantly during the dry season. The local warming impacts of deforestation are well recognized by people living within tropical forest landscapes. Researchers used satellite data to show that deforestation in the Amazon caused substantial warming even up to 100 kilometers away from the location of forest loss. This nonlocal warming increased deforestation-induced warming by a factor of 4.[12]

Clearing forests for industrial agricultural monocultures results in large areas of bare soil. Researchers have found that exposed soil heats up faster than air, further increasing air temperatures. The hot extremes of soil increased faster than air extremes by 1.3°F (0.7°C) per decade in intensity from 1985 to 2020. The researchers identified soil temperature as a key factor in the soil moisture–temperature feedback, stating, "During dry and warm conditions, the energy absorbed by the soil is used to warm the soil, increasing . . . surface air temperatures."[13]

The increase in surface air temperature leads to a higher atmospheric demand for water, increasing soil evaporation. This can further dry and warm the soil, causing hot temperature extremes.[14]

The Vapor Pressure Deficit (VPD)

The total amount of water vapor the air can hold is called the saturated water vapor pressure (SVP) level. When this level is exceeded, the excess water vapor turns into rain. Rising air temperature increases SVP by approximately 7 percent per degree Celsius. When there isn't enough moisture available through soil evaporation and plant transpiration to meet the SVP level, the difference is called a vapor pressure deficit (VPD).

VPD increases if the actual atmospheric water vapor content does not increase by the same amount as SVP. It has a drying effect as it increases the amount of moisture drawn out of the soil through evaporation and from plants through transpiration in order to fix the deficit to meet the SVP requirements.[15]

VPD occurs over land when there is insufficient moisture for the SVP requirements. The increases in SVP (at approximately 7 percent for every degree increase) heightens the drying strength of the atmosphere when there is a VPD. This is the cause of the increase in extreme weather events, such as torrential rain, causing floods and more frequent and longer droughts. Consequently, VPD is a major climate forcing.[16]

Photosynthesis declines when atmospheric VPD increases due to stomata closing. Stomata are the breathing pores in leaves. They take in air and, very critically, CO_2 for photosynthesis. Plants close their stomata when it is too dry to prevent water loss through transpiration. This stops photosynthesis due to not having CO_2 to convert into glucose.

Research shows that VPD, rather than changes in rainfall, substantially influences vegetation growth. VPD notably affects forest mortality and crop yields. Increases in VPD greatly limit land evapotranspiration in many ecosystems because plants close their stomata to stop water loss via transpiration.

This has enormous implications for most forest systems globally, as they produce much of their rainfall via plant transpiration. The Amazon is a good example: Due to forest clearing that heats up the soil and increases the VPD, droughts and high-temperature extremes are increasing.[17]

Satellite observations show that some regions of the planet have been experiencing widespread vegetation greening since the 1980s, primarily due to warmer temperatures and increased CO_2 fertilization. Photosynthesis uses solar energy to combine CO_2 and water to produce glucose, the main building block of life. Plants use glucose to construct cellulose, the fundamental organic compound that forms wood, stems, roots, and leaves. Around 95 percent of a plant's biomass (body) comes from this process. This greening mitigates global warming by increasing CO_2 removal from the atmosphere through photosynthesis to form cellulose and other organic molecules essential for life.

Research into the impact of vegetation greening has shown that it cools the regions where it occurs, which slows down global warming. This cooling effect can partially offset increasing temperatures. The study results highlighted the necessity of considering the effects of vegetation when developing local climate adaptation strategies.[18]

Recent research shows that this greening has stopped in many regions due to the increased VPD drying out land on all continents. VPD is critical in determining plant photosynthesis. Researchers looking at multiple datasets found that a sharp increase in VPD started in the late 1990s. They found that the vegetation greening trend indicated by satellite measurements, evident before the late 1990s, had subsequently stalled or reversed in these regions.

Their research results highlighted that the impacts of VPD on vegetation growth should be adequately considered to assess ecosystem responses to future climate conditions.[19]

Indeed, some researchers have been highly critical of the current climate change models for not including VPD as a major climate forcing. They found that it will become increasingly crucial for vegetation function. A team of scientists from multiple US universities stated, "Our results suggest that failure to consider the limiting role of atmospheric demand in experimental designs, simulation models, and land management strategies will lead to incorrect projections of ecosystem responses to future climate conditions."[20]

Forest and Grassland Fires

Forest fires are another major issue. Most of them are caused by human mismanagement. Fire is the main tool that humans have used to clear and manage ecosystems for over 100,000 years. Damaging forest fires are increasing in frequency, and the acreage burnt is growing. The increase in temperature and VPD due to climate change and the disruption of soil and vegetation hydrology is a major contributor to forest and grassland fires.

Again, VPD is an absolute measure of the difference between the air's water vapor content and saturation value. VPD accurately measures the ability of the atmosphere to extract moisture from the land surface and is critically important in studies of the links between meteorological conditions and forest fires. It has been shown to be closely related to variability in burned forest areas in the western United States.[21] As the climate warms and VPD increases, it will drive even more forest fires. The careful management of forests, farms, and grasslands, especially their hydrology, becomes critical to prevent destructive fires.

Industrial forestry, which manages forests purely for timber products rather than as biodiverse ecosystems, greatly contributes to fires. Practices such as spraying out fire-resistant species like aspen to leave fire-prone species such as pines change forests to make them vulnerable to fires. Planting monocultures of fire-prone trees such as eucalyptus, pines, and acacia instead of regenerating the diverse native ecosystems creates forests that readily burn.

Logging rainforests opens the canopy, letting fire-prone species such as grasses dominate the understory and allowing direct sunlight to dry out the soil. Rainforests rarely burn, as their moisture prevents fires. However, logged forests readily burn. Rainforest trees are very sensitive to fires, so most species are killed, destroying the forest.

On the other hand, traditional management systems such as grazing reduce the grass and flammable bushes. Before forests and their wildlife were severely disrupted by agriculture, industrial forestry, and endocrine-disrupting chemicals, they had a rich species biodiversity that fed on the flammable grasses and underbrush. This biodiversity helped to prevent widespread fires. Traditional pastoralists such as Masai, Fulani, Tuaregs, Toubou, Bedouins, Mongols, Tatars, Sami, Nenets, Chukchis, and Swiss Alpine dairy farmers moved their herds over the landscape to fresh pasture so they never overgrazed and damaged the ecosystems and soils. The traditional systems can be emulated by rotational grazing and browsing livestock such as goats, sheep, and cattle. They can eat all the highly flammable plant species, reducing fire risk.

The Indigenous civilizations in Australia and the plains of North America used cool and limited burns to reduce highly flammable species. The new grass and shoots in the regenerating pastures would attract endemic species such as kangaroos and buffalo. Indigenous nations could effectively create a rotational grazing system, preventing overgrazing and providing their livestock with fresh, highly diverse species.

The rangelands they created, such as savannas and prairies, are incredibly diverse ecosystems with thousands of species.

These North American and Australian Indigenous civilizations thrived for thousands of years until they were destroyed by colonists from Europe who stole their land. Their civilizations had thrived for far longer than sedentary farming communities such as the Egyptians, Romans, Maya, and Khmer. The introduction of fixed grazing along with tillage in sedentary agricultural communities disrupted these rotational systems, severely reducing the ecosystem diversity. Modern industrial agriculture continues to destroy highly biodiverse ecosystems, reducing them to endless vistas of single species such as corn and soy.

Traditional cool-season and storm-season burning systems have been shown to reduce damaging fires. These systems burn small areas to avoid large destructive fires. They are done over months to create a mosaic of fire breaks and regenerated pastures for native animals and livestock.

The so-called "controlled burns" used by various government agencies and fire departments to prevent fires are nothing like these traditional systems. They burn large sections of ecosystems and are far hotter than traditional fires, resulting in severe damage to ecosystems. Also, their ability to prevent large fires is questionable as the amount of vegetation they kill produces more flammable material for future fires.

That's the Bad News; What Happens Next?

Most scientists agree that the world is warming: air and sea temperatures are increasing, ice sheets in Greenland and Antarctica are melting, most of the world's glaciers are shrinking, and sea levels are rising. Solar and planetary cycles, greenhouse gases, and forest and vegetation cover all play a role in climate forcing and contribute to the changing climate. While we cannot change solar and planetary cycles, we can change greenhouse gas emissions and regenerate ecosystems to cool the planet and restore a healthy climate.

The critical issue is that the extra energy in the atmosphere since 1750 (4.1 W/m^2) fuels a changing climate. The surface area of the Earth is 510 trillion square meters. The extra energy of 4.1 W/m^2 means that 2,091 trillion watts of energy have been added to the Earth's atmosphere and oceans since 1750, the start of the Industrial Revolution. This is the equivalent of the energy of millions of atomic bombs affecting our weather.

This extra 2,091 trillion watts of energy are already violently fueling and disrupting our weather systems. The energy is causing weather events to be far more intense. Winter storms can become colder and be pushed further south and north than usual due to this energy, bringing damaging snowstorms and intense floods. Similarly, summer storms, especially hurricanes, tornadoes, tropical lows, and more, are far more intense, with increases in deluging destructive rainfall and floods. Droughts and heat waves are more common, resulting in more crop failures. They are also fueling damaging forest and grass fires that are burning out whole communities and changing regional ecologies due to not allowing time for recovery before subsequent fires.

The frequency and intensity of these events will only worsen exponentially when the world warms to 3.6°F (2°C), the upper limit of the Paris Climate Agreement. We are on track to shoot far past this goal.

Adapting To and Mitigating Climate Change

The dominant approach to climate change is to reduce greenhouse gases, mainly CO_2, by reducing fossil fuel use and increasing the amount of renewable energy. So far, this approach has been a total failure.

Measurements from the Mauna Loa Observatory on Hawaii's Big Island averaged 424 ppm in May 2023, up 3 ppm from the May 2022 average, according to NOAA data.[22]

Four hundred one ppm CO_2 was recorded in 2015, the year the United Nations Paris Agreement was adopted by 196 countries as an international treaty on climate change.[23] The goal was to limit global warming to below 3.6°F (2°C), preferably to 2.7°F (1.5°C).[24] That means that CO_2 emissions have been increasing by about 2.87 ppm per year since 2015. Emissions increased at 2 ppm per year in the decade *before* the Paris Agreement, so the emissions rate is rapidly accelerating.

The 2014 Intergovernmental Panel on Climate Change (IPCC) Mitigation of Climate Change report stated that greenhouse gas concentrations should be limited to 430 ppm carbon dioxide equivalents (CO_2-eq) to limit warming to 1.5°C and to 450 ppm CO_2-eq to limit warming to 2°C.[25] According to the IPCC report, we have overshot the 1.5°C target. The IPCC estimated that 430 ppm CO_2-eq was reached in 2011.[26]

CO_2-eq combines carbon dioxide and other anthropogenic greenhouse gases, such as methane, nitrous oxide, and halocarbons. Their warming potentials are expressed in units equivalent to CO_2. Combining all these greenhouse gases allows us to better understand the levels of gases contributing to climate change.

The IPCC stated that additional mitigation is needed; otherwise, the temperature will overshoot the Paris Agreement targets and increase the global mean surface temperature in 2100 from 6.7°F to 8.6°F (3.7°C to 4.8°C) compared to preindustrial levels.[27] Since the targets have already been overshot, and the emissions rate is accelerating, reducing emissions and transitioning to renewable energy are no longer sufficient to stop severe warming.

The IPCC states that the only way to limit global warming to 2.7°F (1.5°C) is to achieve net negative emissions (reverse emissions) by using carbon dioxide removal (CDR) to draw down CO_2 from the atmosphere. The IPCC has advocated for CDR technologies such as regenerating natural ecosystems, soil carbon sequestration, and carbon capture and storage.[28] Indeed, there are many so-called solutions on the table, and it can be hard to separate the wheat from the chaff.

So-Called Solutions—and Their Problems

So-called solutions to our climate crisis include renewable energy (notably solar and wind), nuclear power, carbon capture and storage, and geoengineering schemes, such as blocking sunlight. Major media campaigns and government legislation are aimed at reducing emissions through closing down farms and reducing animal production, especially cattle and sheep production.

These solutions come with a set of major environmental problems that are primarily ignored, resulting in community conflict, farmer protests, and increasing opposition to the rollout of renewable energy systems and transmission lines in rural areas.

Renewable Energy

The rollout of wind turbines and large-scale solar farms is generating a lot of negative sentiment. Wind turbines are making negative headlines because they produce low-frequency noise, cause a range of health issues for neighbors, lead to the death of birds (especially endangered and rare species such as eagles), necessitate the clearing of ecosystems, and ruin the aesthetics of natural environments by turning them into industrial landscapes.

There are numerous battles between communities and some environmental groups to stop the construction of more wind turbine farms. For example, where I live, thousands of acres of valuable tropical forests that are home to endangered species are being cleared for wind farms. Over 16 million trees have been cut down in Scotland for wind farms. Parts of the environmental movement are silent on the environmental damage they cause, which is causing rifts over best methods for responding to climate change.

Similar issues are occurring with solar electric panels, as their installation can lead to loss of farmland needed for food production and the clearing of natural ecosystems. There are proposals to build solar farms on thousands of square miles to meet "clean" energy needs. If they are built, the scale of clearing will be one of the most significant environmental disasters on the planet. Communities are unhappy with the loss of their farm landscapes, the toxic metals that leach into the environment, and the radiation caused by being in proximity to solar farms. They are protesting the thousands of miles of new high-voltage transmission lines proposed or constructed across their farms and landscapes to connect the new renewable systems to the existing power grids.

Renewable energy systems require multiple mines to provide the metals and other compounds needed for manufacturing and construction. The Hongana Manyawa people of Halmahera, some of the last uncontacted tribes in Indonesia, are fighting to save their rainforests and traditional cultures from genocidal destruction caused by nickel mining for batteries and solar cells.

There are large-scale plans to log Halmahera's unique rainforest areas and mine them for nickel. The Hongana Manyawa people are these valuable tropical ecosystems' traditional owners and protectors. The forests provide their livelihood, and their destruction will cause the genocide of some of Asia's last uncontacted rainforest communities. Companies, including Tesla, are investing billions of dollars into Indonesia's plan to become a significant nickel producer for the electric car battery market. French, German, Indonesian, and Chinese companies are involved in mining in Halmahera.

Cobalt mining in the Congo for lithium batteries and electric cars exemplifies the worst cruelty, exploitation, and oppression of workers, especially children, in the world today. Described as modern-day slavery using cruel child labor, the miners work in subhuman, grinding, degrading conditions for a few dollars a day.

Cobalt mining has decimated the Congo's landscape. The rainforests are disappearing, and the local climate has deteriorated because millions of trees have been destroyed. Cobalt is toxic to touch and breathe. Mining has caused the air to become hazy with toxic dust and contaminated the water with toxic effluents. Hundreds of thousands of poor Congolese, including mothers with babies strapped to their backs, constantly touch, breathe, and drink toxic cobalt.

Compared to traditional energy systems, the short life cycles of renewables—around 20 years—result in toxic waste disposal problems. Old wind turbines are buried or left in piles on the ground. The disposal of used solar cells creates similar degenerative environmental waste problems. Toxic heavy metals and forever chemicals leach into the environment. Because they are rarely recycled, renewable energy systems are not renewable. The recycling of renewable energy systems must be addressed instead of allowing them to pollute the environment. This is critically important as all the current renewable systems will need to be replaced before 2050 when the world is supposed to achieve net zero emissions.

Systems such as rooftop solar panels and wind turbines with batteries connected through local microgrids can effectively ensure household energy needs. These standalone systems do not need industrial-scale solar and wind farms, nor high-voltage transmission lines. These systems must be concentrated in the towns and cities that consume the power rather than in rural landscapes, which require industrialization and the transportation of power over thousands of miles of high-voltage power lines.

Biofuels

Biofuels, on the whole, are highly problematic. Large areas of food-producing farmland are used to fuel cars, trucks, and airplanes rather than feed people. Worse still, vast areas of tropical forests have been and are still being cleared for biofuels, such as palm oil and GMO maize. They are not greenhouse gas–negative because fossil fuels are used in their production. Burning them for fuel produces CO_2, the main greenhouse gas.

Furthermore, the synthetic nitrogen fertilizers used to grow biofuels are produced using fossil fuels, and their use causes nitrous oxide emissions, a greenhouse gas much more potent than CO_2. Other studies show that using these fertilizers causes the oxidation of soil organic matter, releasing thousands of tons of CO_2 into the atmosphere. Quality life-cycle assessments of all the parameters used to produce biofuels show that they contribute to atmospheric greenhouse gases. They constitute a significant part of the problem, not a solution.

Nuclear Energy

Nuclear fuels are touted as a clean and reliable source of energy. But nuclear power is now far more expensive to produce than power from sources like solar and wind. The nuclear fuel cycle also causes massive environmental problems that have never been solved despite endless promises. The mining and processing of uranium causes long-term environmental damage that continues for centuries and significantly contributes to greenhouse gas production. There is no viable way to dispose of spent nuclear fuel rods and cooling water. Most nuclear waste is still stored in unsafe temporary sites.

Nuclear fuel rods begin to emit highly lethal gamma rays after one to two years in the reactor core. Spent fuel rods are stored in pools of water for decades as they cool down and are unsafe to approach unless shielded by many feet of water. This storage water becomes radioactive, and so far, there is no safe way to store it. In August 2023, the Japanese government began releasing hundreds of tons of radioactive cooling water from the Fukushima reactors into the Pacific Ocean because they could not store it safely. They say it is safe, but many experts argue that this radioactivity will bioaccumulate in marine food chains, causing long-term health and reproductive problems for multiple species and people who eat seafood across the Pacific, including North America. Consequently, China banned the import of all Japanese seafood.

Ten years after removal from a reactor, the toxic radiation from a spent nuclear fuel rod exceeds 10,000 rem/hour. The fatal one-time exposure dose for humans is 500 rem. The decay of nuclear fuel is measured in half-lives. For example, a standard estimation of the half-life of many spent fuel rods is 24,000 years. This means that after 24,000 years, half of the fuel is left. After 48,000 years, one quarter is left. After 72,000 years, one-eighth is left, and at 96,000 years, one-sixteenth still exists. Uranium-233 from fuel rods using thorium has a half-life of 159,200 years. After 636,800 years, one-sixteenth remains, polluting the environment. The considerable accumulation of thousands of tons of spent fuel and other highly radioactive waste means that even after millions of years, the toxic radioactive residues are considerable.

While most spent radioactive fuel is stored for decades in temporary unsafe pools of water, there are proposals for longer-term sites such as salt mines and deep underground granite caverns. Most of these proposals have been canceled due to technical problems and community opposition because they pose risks of leaching into the wider environment. This is already happening where spent fuel has been stored in salt mines in Europe. There is zero credible science to show that any storage system can be safe for the millions of years needed before the radioactivity has decayed to normal background levels that are considered safe. Scientists cannot predict that the storage caverns will not be damaged by earthquakes, leaching, or other events, and cause the release of these lethal poisons into the environment.

The meltdown risk continues, with Chernobyl and Fukushima still causing problems decades later. Both accidents caused massive spikes in cancer rates in communities exposed to the clouds of radioactivity. The issue of wars and terrorism causing meltdowns is genuine, with Europe's largest nuclear reactor subject to shelling in the war between Russia and Ukraine. Decommissioning nuclear power plants takes decades and billions of dollars and requires extensive use of greenhouse gas–producing fossil fuels. The radioactive parts need to be disposed of, and like the spent fuel rods and cooling water, there is no proven safe way to do this. The bottom line is that nuclear power is too dangerous and too expensive to be considered a positive solution to climate change.

Carbon Capture and Storage (CCS)

Carbon capture and storage (CCS) is promoted as a carbon drawdown technology for reducing greenhouse gas emissions.

A report from the IPCC stated that CCS is essential to reach net zero emissions by 2050.[29]

A review of all the major carbon capture projects found that over 70 percent were used for enhanced oil recovery. Oil and gas companies use the captured CO_2 to pump more oil and gas out of depleted wells, producing more greenhouse gas emissions. The study reviewed the 13 large-scale CCS projects currently in existence worldwide. It found that seven underperformed, and one was questionable. Nearly 90 percent of the proposed CCS capacity in the power sector failed at the implementation stage or was suspended early. Only two projects in the gas processing sector demonstrated some success.[30]

Many CCS projects propose transporting CO_2 in pipelines over hundreds of miles under extremely high pressures to maintain a supercritical fluid state. The high pressures can rupture the pipes over long distances, allowing CO_2 to escape before the flow can be stopped. The odorless, invisible gas can travel long distances in plumes, suffocating humans and animals without warning, making the pipelines an existential threat to the communities they pass.

Most of the proposed storage proposals, such as in disused mines and wells, aquifers, and underground caverns, have no long-term scientific studies to show that the CO_2 will not escape back into the atmosphere.

The report shows that billions of dollars have been invested into this sector with few results. The carbon capture sector is still a net emitter of greenhouse gases. And yet billions more dollars are still being budgeted for industrial CCS projects.

Geoengineering

There are many geoengineering solutions being touted, such as spraying sulfur dioxide from planes or using mirrors in space to block the sun. These are potentially the most dangerous and damaging of all prospective "solutions." Blocking the sun would adversely affect agricultural production because such production relies on solar energy to power photosynthesis. Similarly, such proposals would negatively affect *all* ecosystems, since all ecosystems require sunlight. The effects on long-term weather and climate are entirely unknown and impossible to predict by current weather and climate system models.

These so-called proposals are all flailing and failing. They must either be abandoned completely or modified to become more effective and less destructive. There are ways of improving some of these strategies, and that must become a priority, especially when scaling them up. Unfortunately, this is not happening at the moment. The leading environmental NGOs must take full responsibility for ignoring the collateral damage caused by these strategies and failing to insist upon solutions that regenerate, rather than degenerate, the planet. As it is, these "solutions" are destined to leave a lasting legacy of destroyed ecosystems and long-term toxic pollution.

So far, mainstream solutions have not made any difference to the emissions rate, which continues to increase. It will be necessary to scale these technologies up to levels 100 times greater than currently exist to replace fossil fuels. This will cause a massive loss of ecosystems and increase environmental damage and community conflicts. Currently, they are not solutions. The growing evidence shows they are the next generation of significant environmental problems and sources of community conflicts.

CHAPTER THREE

The Promise and Potential of Regenerative Agriculture

Trillions of dollars are now being spent on carbon capture and storage, nuclear power, biofuels, solar cells, and wind turbines. Instead of going down, however, the rate of CO_2 emissions *increased* from 2 ppm per year in the decade leading up to the Paris Agreement in 2015 to 2.87 ppm per year, setting a new record of 424 ppm in May 2023. It is evident that the current solutions are failing.

Proposals to reduce emissions frequently involve closing down farms and reducing animal production and meat and dairy consumption. Governments such as New Zealand are proposing taxing ruminant production to minimize production in order to reduce methane output. At the time of writing, the Netherlands plans to forcibly close down farms and remove their land from primary production, and Ireland intends to force farmers to kill 200,000 cows.

Industrial agriculture poses an existential threat to the planet through deforestation, biodiversity loss, toxic poisons, soil loss, eutrophication, and greenhouse gas emissions. In fact, industrial agriculture is directly and indirectly responsible for 30 to 50 percent of greenhouse gas emissions.[1] Ruminant livestock systems, in particular, are a significant source of agriculture's anthropogenic greenhouse gases. The United Nations Food and Agriculture Organization's (FAO) research has shown that cattle, buffalos, goats, and sheep for meat and milk generated 5.8 gigatons of CO_2-eq in 2010.[2] (A gigaton is 1 billion tons. A metric ton and a US ton are almost the same; for the sake of simplicity, they are regarded as the same in this book.)

There is a huge difference, however, between high-intensity concentrated animal feeding operations (CAFOs) and extensive pasture-based grazing systems on the world's 8.4 billion acres (3.35 billion hectares) of permanent pastures. This difference persists not only in terms of carbon emissions and sequestration, but also in terms of all the other benefits and detriments, such as impact on soil organic matter (SOM), biodiversity, human health, and local economies. (See appendix for details and methodology on calculations.)

This delineation is also crucial, more broadly, between industrial agriculture and regenerative agriculture—including agroecology, agroforestry, permaculture, and organic agricultural systems. We simply cannot compare apples to oranges when evaluating the benefits and drawbacks of these vastly different ways of raising food and managing land.

The good news is that multiple examples of regenerative agricultural systems are being scaled up globally to regenerate biodiversity, communities, climate, and our health. (I've dedicated chapter 4 to four of these inspiring examples in Ethiopia, the Philippines, India, and China.) This is not a complex, impractical academic dream or a long-winded proposal by bureaucrats. It is happening now with minimal governmental, academic, or industry support. Regenerative agriculture is a grassroots revolution pioneered mainly by farmers and ranchers.

What Is Regenerative Agriculture?

Most people are familiar with the (now overused) term "sustainable," which characterizes practices that maintain resources and the environment without degrading them. In an already degraded system, however, it is necessary to do more than sustain the status quo: We must set it on the path to healing—or, regeneration. Regenerative systems, therefore, not only sustain existing systems, but improve them. For agriculture and environmental management, this means improving the soil, plants, animal welfare, human health, and human communities. Significantly, it includes practices that increase the resilience to climate change–induced weather impacts and reduce production costs while ensuring reasonable returns to farmers and land managers. This ability to improve climate change adaptation and economic viability is essential for the future of agriculture and ecosystem regeneration.

The opposite of regenerative is degenerative. This is an essential distinction in terms of agricultural activities. Agricultural systems that use degenerative practices and inputs that damage the environment, soil, health, genes, and communities and involve animal cruelty are not regenerative, and never will be (no matter how much they're greenwashed). The use of synthetic toxic pesticides, synthetic water-soluble fertilizers, and GMOs; the clearing of ecosystems; confined animal feeding operations; exploitive marketing and wage systems; and destructive tillage systems are examples of degenerative practices.

Regenerative agriculture, on the other hand, covers the many farming systems that use techniques such as longer rotations, cover crops, green manures, legumes, compost, and organic fertilizers. More specifically, regenerative agriculture maximizes the photosynthesis of plants to capture CO_2 and sequester it as soil organic matter. It can be applied to all agricultural sectors, including cropping, grazing, and perennial horticulture. It is being used as an umbrella term for a range of systems such as organic agriculture, agroforestry, agroecology, permaculture, adaptive multi-paddock grazing, silvopasture, analog forestry, syntropic farming, pasture cropping, and other agricultural systems that can increase SOM.[3]

Regeneration International asserts that to heal our planet, all agricultural systems should be regenerative, organic, and based on the science of agroecology. Farmers can determine acceptable and regenerative practices using IFOAM - Organics International's Four Principles of Organic Agriculture. These principles are:

1. **Health.** Organic agriculture should sustain and enhance the health of soil, plant, animal, human, and the planet as one and indivisible.
2. **Ecology.** Organic agriculture should be based on living ecological systems and cycles, work with them, emulate them, and help sustain them.
3. **Fairness.** Organic agriculture should build on relationships that ensure fairness in the familiar environment and life opportunities.
4. **Care.** Organic agriculture should be managed in a precautionary and responsible manner to protect the health and well-being of current and future generations and the environment.

The Science Behind Regenerative Agriculture

Carbon dioxide, the main greenhouse gas, is the basis of all life on our planet. Plants, through photosynthesis, use solar energy to turn carbon dioxide and water into glucose. Glucose is the basis of the food system for most of life. It is a primary energy source of the cells of most living organisms, including plants and animals. It is the basis of the fuels that power mitochondria—the engines inside the cells of nearly every organism on Earth, including us.

Glucose molecules are transformed to build thousands of types of sugars, such as sucrose (cane sugar), fructose (fruit sugar), lactose (milk sugar), maltose (malt sugar), and many others. Glucose molecules can be combined in long chains to form cellulose. Cellulose is the basis of wood, leaves, stems, and paper. Glucose molecules are also transformed into the carbohydrates found in flour, bread, and staples such as rice, wheat, corn, potatoes, cassava, taro, and so on. These carbohydrates can be transformed into hydrocarbons—oils and fats.

Glucose molecules are also transformed, with the addition of nitrogen and sometimes sulfur, into amino acids—the basis of proteins, DNA, hormones, nerves, muscles, eyes, the brain, bone marrow, blood cells, stem cells, and all the tissues that make up our bodies and the bodies of all living organisms. These organic molecules make life, and they all come from glucose.

Between 95 and 98 percent of a plant's biomass (its "body") comes from water and carbon dioxide via the glucose produced through photosynthesis. Only 2 to 5 percent comes from soil minerals. There is a lack of understanding of the fundamental basis of the organic molecules that make up life in standard agronomy, the science used by industrial agriculture. Standard agronomy has taken a reductionist approach by focusing on dead chemicals and minerals and has ignored the biological science that contributes to over 95 percent of living organisms. CO_2 and water are the most essential components of all life on our planet. CO_2, instead of being seen as a problem, needs to be seen as a powerful asset to regenerate our planet's biodiversity, climate, and health. The key is the correct agricultural management systems.

Ignoring this fact has led to numerous damaging agricultural practices that increase production costs by destroying soils, increasing pests and diseases, poisoning our food and environment, stopping the natural regeneration of soil biology and fertility, lowering water capture and retention, and reducing yields in the medium and longer terms. It has turned agriculture into a significant contributor to climate forcing through greenhouse gases, deforestation, bare soil, and damaged hydrology.

The overwhelming majority of life on Earth is dependent on the products of photosynthesis either directly, as with plants, or indirectly, as with microorganisms and animals. Plants and microorganisms that photosynthesize provide the basis of most life on this planet. Animals are directly or indirectly dependent on plants for their nutrition. Carnivores need to eat herbivores, animals that eat plants. Without plants, animals, including humans, would starve. It is the same with most microorganisms, which depend on living and dead plants for their primary energy sources: glucose, oils, fats, and amino acids, the molecules of life.

Many microorganisms live off dead plants, breaking down and recycling cellulose, amino acids, and carbohydrates into the necessary nutrients to grow and reproduce. However, the greatest concentration of microorganisms lives in symbiosis with living plants.

Broadly speaking, regenerative systems focus on maximizing the number of living plants. Only living plants, not dead plants, produce the molecules of life. These molecules of life are the basis of soil organic matter. The vast majority of soil carbon comes from the CO_2 in the air captured through photosynthesis.

The Origins of Regenerative Agriculture

Regenerative agriculture has its roots in the organic, permaculture, agroecology, and biodynamic movements. The publication of Rudolf Steiner's agriculture lectures, which he gave in Germany in 1924, started the biodynamic movement, the first formal movement of a new farming paradigm. Then in 1940 Sir Albert Howard published *An Agricultural Testament*, popularizing the concept of soil health and the primary importance of soil organic matter. Howard had spent much of his time in India where he learned composting from Indian farmers. He pioneered efficient forms of composting that achieved high yields of healthy plants. Howard stated: "The health of soil, plant, animal, and man is one and indivisible."

Howard's ideas in turn had an enormous influence on J. I. Rodale, who began publishing *Organic Farming and Gardening* magazine in 1942 to promote his methods based on the widespread use of recycling organic matter through composting, green manuring, and mulching. Rodale was the person who popularized the term "organic farming" worldwide and is considered the father of the modern organic farming movement. Rodale, being a publisher, simplified Howard's statement to: "Healthy Soil = Healthy Plants = Healthy People."[4]

The publication of *Silent Spring* in 1962 by Rachel Carson significantly raised public awareness about the dangers of pesticides used in farming. *Silent Spring* created considerable controversy and massive concern about the chemical residues in foods and the environment. The resulting public pressure saw the strengthening of pesticide regulations and, most importantly, the beginning of the consumer movement that demanded food grown without toxic chemicals.

It also saw the beginning of environmental awareness around farming. It gave rise to several "whole-system" approaches that fit the regenerative paradigm, such as that described in *The One-Straw Revolution* by Japanese farmer Masanobu Fukuoka, which was published in English in 1978 and quickly became one of organic agriculture's most influential books. Fukuoka's "natural farming" methods were based on observing how nature works and designing the system so that nature did the work for you. He was one of the pioneers of organic no-till grain systems that did not use herbicides. These systems are easily applied to small farms.

Independently of Fukuoka, two Australian researchers, Bill Mollison and David Holmgren, published a book called *Permaculture* in 1979. Permaculture was a shortened word for "permanent agriculture," a concept first put forward by F. H. King in the book *Farmers of Forty Centuries* in 1911.

Permaculture is a comprehensive whole-system approach to designing integrated systems that include cropping and animals. Permaculture works with ecology, horticulture, vertically stacked production systems, animals, water, architecture, energy efficiency, and numerous other concepts. The ideal is to start with a vacant block of land and design the new system based on the specifics of that block, its climate, its topography, and other attributes. Each permaculture farm is unique because of this.

In 1979, three books on agroecology were published. Steven Gliessman published *Agroecosistemas y tecnologia agricola tradicional*, George Cox and Michael Atkins published *Agricultural Ecology: An Analysis of World Food Production Systems*, and Robert Hart published *Agroecosistemas: conceptos básicos*. These books ushered in the beginning of the agroecology movement, which is defined as "both a science and a set of practices. As a science, agroecology consists of the application of ecological science to the study, design and management of sustainable agroecosystems. . . . This implies the diversification of farms in order to promote beneficial biological interactions and synergies among the components of the agroecosystem so that these may allow for the regeneration of soil fertility, and maintain productivity and crop protection."[5]

In other words, agroecology is both a science and a movement. It has powerful adherents in Latin America and Africa and, to a lesser extent, Asia, Europe, North America, and Australasia. Many of these movements are highly political, focusing not just on farming production systems; they also concentrate on the rights of farmers to have a fair standard of living and the rights of communities to have food sovereignty. In addition, agroecology is now taught and researched in many universities on every arable continent, and is gaining considerable credibility, with a growing number of peer-reviewed scientific publications.

Agroforestry and Its Offshoots

There are many specific farming techniques that can fall under the umbrella of regenerative agriculture, including agroforestry, AMP grazing, pasture cropping, and more. Although the specific practices may differ—and, in practice, they often overlap—they share in common the use of highly biodiverse perennial plants and animals in farming systems rather than seasonal monocultures, often scaled to create new permanent agroecosystems that regenerate landscapes and their communities.

For example, the agave agroforestry system is designed to regenerate degraded arid and semiarid regions. Numerous other agroforestry systems have been developed for all climates and soil types. These systems continue to expand and diversify into many forms, such as row or alley cropping, where the annual crop, such as wheat, sorghum, or maize, is cultivated in strips between rows of trees. Many of these are "fertilizer trees" because they fix nitrogen, increase organic matter, and provide other nutrients for the cash crop. Their leaves and branches can be harvested and left to biodegrade into organic matter to feed the soil microbiome and improve soil fertility.

To take another example, tropical ecologist and researcher Mike Hands developed inga alley cropping in Honduras. Ingas are called ice cream bean trees because the pulp around the seeds of several species tastes like ice cream. They are also a nitrogen-fixing legume and important pioneer species in the tropical Americas, contributing to ecosystem regeneration in multiple ways. I have observed this in South America and have seen how vital the fruit pods are as income for the local communities. I have several species of inga on my own farm in Australia and am impressed with their fast growth and early fruit production.

Inga alley cropping maintains soil fertility and good harvests year after year, allowing families to gain long-term food security on one piece of land. Once the inga alleys have developed to the stage where they shade out weeds, they are pruned to chest height. The smaller branches and leaves are used as mulch and organic matter to build soil fertility as they decay. The larger branches are used as firewood, allowing families to obtain all the wood they need for cooking instead of cutting down trees in nearby forests, ending one of the causes of deforestation.[6]

The cash crop is planted in the mulch within the alleys. The ingas grow until the next planting season, and the cycle repeats. This system ensures reliable harvests yearly on the same plot of land using minimal labor. Compared to cleared fields, the shading from the inga trees cools the landscape, conserves soil moisture, and restores the hydrology. Like the agave agroforestry system developed by the Flores González brothers in Mexico, the inga alley cropping system exemplifies the compounding ecosystem benefits of agroforestry and regenerative agriculture.

Perennial agroforestry food systems can also be designed with rows of fruit trees and diverse understory plants that exude the organic molecules of life from their roots to feed the soil microbiome and generate soil fertility. Stratification takes advantage of different plants sizes to maximize the capture of sunlight and thereby increase the efficiency of photosynthesis. Smaller fruiting species such as bananas, papayas, and berries can be grown in between larger fruiting tree species. Flowering plants that host beneficial insects that are natural enemies of pests can be added for pest and disease control. Other plants can be grown as vegetables, tubers, and herbs for feed, such as taro, yams, turmeric, and ginger. These understory plants can be managed by rotationally grazing them with livestock to make highly productive systems that generate most of their fertility and pest and disease control on farms for minimal costs.

To take it yet another step further, silvopasture systems graze animals under trees rather than in cleared fields. In fact, the agave agroforestry system is an example of silvopasture as well. The trees provide shade and shelter for the animals and forage in drought. This also is what I have done with my own farm. These layered designs are among the most productive agricultural systems.

Syntropic farming was pioneered by Ernst Gotsch in Brazil in the 1980s. It is an evolving example of an agroforestry alley cropping system. This agroforestry system provides yields at all stages of plant growth and succession and generates fertility by creating a productive forest that imitates the structure and function of native forests.

Syntropic farming is based on the principle of keeping the soil covered with living plants and maximizing photosynthesis through the stratification, synchronization, and natural succession of a high diversity of plants, from small vegetables to tall trees. The plants are pruned and harvested to provide a continuous supply of organic matter to increase soil fertility. Syntropic farming is gaining popularity, with a new generation considering the system an advanced form of agroforestry for its potential to combine high productivity with large-scale regeneration.

To take a final example, analog forestry was developed by Dr. Ranil Senanayake in Sri Lanka in 1981 as an alternative to monocultures. They are designed to mimic indigenous forests' structural and functional aspects. Analog forests are designed to provide economic benefits and may comprise a mix of natural and exotic species needed for food and income. A species' contribution to the forests' structure and function is the determining factor in its use. The forests comprise edible fruit trees, vines, shrubs, herbs, and vegetables grown in sequences analogous to a region's natural forest ecosystem.

Holistic Planned Grazing

Zimbabwean ecologist and livestock farmer Allan Savory is one of the pioneers of regenerative agriculture. In the early 1980s, Allan and Robert Rodale collaborated to promote Robert's term "regenerative organic agriculture" for agricultural systems that improve the resources they use rather than destroying or depleting them. After Robert's passing, Allan continued to work on a definition.

Regenerative agriculture is the production of food and fiber from the biological life of the world's land and waters through managing simultaneously the indivisible complexity of human organizations, economy, and nature to sustain all businesses, economies, and civilization.

Allan popularized Holistic Planned Grazing after his involvement as a young man in a program for the mass culling of wild animals. When he observed that the culling program did not reverse land degradation as intended, Allan began to question whether the prevailing paradigm—that too many animals cause overgrazing and degenerate ecosystems—was correct.

Allan's quest for solutions was influenced by many scientists, in particular the work of French agronomist André Voisin, who discovered that overgrazing resulted from the duration of plant exposure to animals, not animal population size. After trying Voisin's rational grazing in Zimbabwe, where rain is only received about four months of the year, and seeing it fail, Allan realized a more sophisticated planning process was needed to address the greater complexity of erratic rainfall. He developed the Holistic Planned Grazing process that is used today. This process, which uses livestock grazing as a tool to regenerate biodiversity, has proven consistently successful on every continent with arable land for over half a century.

Allan realized that this was the solution to regenerating rangelands. While overgrazing was caused by letting animals graze for too long and returning them before the plants had time to recover, many animals grazing briefly, provided that the vegetation had enough time to recover, mimicked natural grazing by herding animals and increased biodiversity. Even a low number of animals that continuously eat their preferred species can kill plants because they never have the opportunity to recover.

Allan developed the Holistic Management framework, detailed in his book *Holistic Management*, as a way for people to manage the complexity of various types of food and farming operations. Others began promoting various adaptations of Holistic Planned Grazing, such as rotational grazing, cell grazing, mob grazing, and adaptive multi-paddock (AMP) grazing. Unfortunately, Allan hasn't been given the full credit and recognition he deserves as the founder of this global grazing movement.

Allan has repeatedly stated that scaling up holistic management of grasslands can sequester enough CO_2 to reverse climate change. He has been criticized by academics entrenched in industrial agriculture paradigms, who say this is impossible. Emerging peer-reviewed publications show that Allan is correct. Research by Richard Teague and his colleagues shows that changing livestock systems can significantly increase soil organic carbon (SOC) levels. They achieved an average of 11 tons of CO_2-eq per hectare per year (11,000 pounds per acre), which, if scaled up across grazing lands, would sequester 37 gigatons (Gt) annually, resulting in reverse emissions.[7]

In a later study, researchers found similar results and recommended the widespread adoption of regenerative agriculture practices not just for increasing SOC; they also found considerable ecological and biodiversity benefits. Specifically, the researchers found that:

> *Incorporating forages and ruminants into regeneratively managed agroecosystems can elevate soil organic C, improve soil ecological function by minimizing the damage of tillage and inorganic fertilizers and biocides, and enhance biodiversity and wildlife habitat. We conclude that to ensure long-term sustainability and ecological resilience of agroecosystems, agricultural production should be guided by policies and regenerative management protocols that include ruminant grazing.*[8]

Research has also demonstrated that changing practices can rapidly increase SOC. Tong Wang and his colleagues found that when poor management lowers SOC stock over time, transitioning to an improved regenerative practice will increase SOC stock at a higher rate.[9] Researchers using regenerative grazing practices in the southeastern United States sequestered 29.36 metric tons of CO_2-eq per hectare per year. Significantly, the authors gave other examples from research worldwide that achieved similar SOC sequestration levels through regenerative grazing. Hence, the results of this research paper are not an isolated outlier.[10] If these regenerative grazing practices were implemented on the world's 8.4 billion acres of permanent pastures, it would sequester 98.6 gigatons of CO_2 per year, significantly more than the 28 gigatons of CO_2-eq currently emitted annually. This would start to reverse climate change and regenerate the planet's ecosystems.

Allan founded the Savory Institute, now headquartered in Denver, Colorado; the Africa Centre for Holistic Management near Victoria Falls, Zimbabwe; and Holistic Management International, headquartered in Albuquerque, New Mexico. These organizations work with ranchers and farmers worldwide to scale up holistic management on every continent. As of this writing, there are 54 "Savory Hubs" in 30 countries with 203 accredited professionals who have trained 15,755 land managers on 55 million acres (22 million hectares) of land.

Zimbabwean Precious Phiri is a graduate who now serves as a training coordinator for the Savory Institute's Africa Centre for Holistic Management. She lives in Victoria Falls and specializes in training workshops on Holistic Planned Grazing and agroecology for traditional communities in her region and southern and eastern Africa.

In 2023, I visited the Africa Centre for Holistic Management with Precious and conservancy ranch manager Etienne Oosthuizen. Allan originally owned the land, dedicated it to wildlife management, and donated it to the local chiefs as trustees to benefit the people of Zimbabwe and Africa. When Allan bought the ranch in the 1970s, it was severely degraded, with eroded soil on which the previous owner had run 100 head of cattle. Instead of destocking, Allan started regenerating his ranch by managing it holistically. All the vegetation increased within the first year as cattle were used to restore habitats and reverse desertification. By 2009, the ranch had a full cover of grass species instead of degraded, bare soil.

By the time I visited in 2023, the Africa Centre for Holistic Management supported 400 cattle and had so much grass, it needed many more to graze and ensure the grass continued to regenerate. On Allan's advice and coaching, they are building up as rapidly as possible to 1,000 cattle just to keep pace with the increased grass production.

The trees and shrubs increased, creating a beautiful mosaic of ecosystems supporting a wide diversity of species. This, in turn, has provided the food sources and shelter for the wildlife, which has also increased. The conservancy ranch now supports many elephants, buffalo, and other wildlife species. I saw kudu, impalas, baboons, and other species there. The river on the conservancy ranch now flows perennially rather than seasonally in normal years. During my visit, it was full of waterlilies, fish, and an abundance of biodiversity, and bordered by tall, healthy trees, despite twenty consecutive poor rainfall years.

This was a total contrast to the nearby communal lands and national park ranches, which had bare ground and limited biodiversity. Their creeks and rivers had short, seasonal muddy flows. I tell Allan's critics that not only are there solid, scientific peer-reviewed papers to prove that he is correct, but that they should also "ground-truth" his claims and visit his ranch. Seeing is believing.

While the output of methane and other greenhouse gases is considerable for concentrated animal feeding operations (CAFOs) and intensive industrial livestock production systems, this is not true for regenerative grazing livestock practices on pasture. Many quality studies show that these practices sequester more greenhouse gases than they emit. They are greenhouse gas–negative.

In ranch ecosystems, much of the methane emitted by animals on pasture is degraded by soil- and water-based methanotrophic (methane-eating) microorganisms. These organisms do not exist in CAFOs and intensive livestock systems, so 100 percent of their emissions go into the atmosphere.[11] Furthermore, methane is a short-lived greenhouse gas with a half-life of 12 years. It decays into CO_2. This CO_2 is sequestered into the soil by photosynthesis in correctly managed grazing systems. This does not happen in CAFOs and industrial animal production systems.

A meta-review published in the *Journal of Soil and Water Conservation* found that transitioning to regeneratively managed ruminant grazing practices can result in more sequestration than emissions, turning ruminant agriculture from a significant emitter to a major mitigator of greenhouse gas emissions. The researchers found that a permanent cover of forage plants is highly effective in reducing soil erosion and that ruminants consuming only grazed forages under appropriate management result in more CO_2 sequestration than emissions.[12]

Most studies looking at the emissions from livestock systems do not factor in the SOM sequestration levels that can result from different livestock management systems. Researchers doing life cycle analysis comparing different livestock management practices found that converting to AMP grazing resulted in significant levels of SOM sequestration, net carbon–negative livestock management, and the best CO_2 mitigation option.[13]

Pasture Cropping

Pasture cropping is an innovative regenerative agriculture system in which the crop is planted in a perennial pasture instead of in bare soil. There is no need to plow out the pasture species or kill them with herbicides before planting the cash crop. It was first developed in Australia by Colin Seis, with whom I have met and discussed his system several times. The principle is based on the ecological fact that many annual plants will grow in perennial systems. The key is to adapt this principle to the appropriate management systems for specific crops and climates. Pasture cropping can be used on permanent pastures and cropping lands.[14]

Neils Olsen, a good friend of mine, further innovated pasture cropping. He developed equipment that combines cultivation, mulching, aeration, and mixed species seeding into narrow tilled strips in the perennial pasture in one pass. The field is grazed down or mulched before planting to reduce competition with the cash crop.

I have visited and stayed on Neils's farm several times in different seasons to observe the results of his system. The results are impressive, with a substantial difference in soil organic matter, open structure, and moisture content compared to the neighbors. The winter cover crop of multiple forage species grows several feet tall and provides high-quality feed for hundreds of cattle. The neighbors, in contrast, had barely any grass because it did not grow well in winter and was grazed down to less than a quarter of an inch. In many areas in the district, the fields were bare ground. Their livestock struggled to get enough feed in winter, whereas Neils's livestock thrived, grew, and had healthy, glowing coats. Neils has substantially increased his stock-carrying capacity, and his production costs have decreased since he does not need to incorporate expensive synthetic nitrogen fertilizers to stimulate pasture growth or expensive herbicides to kill weeds.

Pasture cropping is an excellent system for increasing SOM. Olsen was paid for sequestering 11 tons of CO_2-eq per hectare per year under the Australian Carbon Credit Unit Scheme in 2019. He was paid for 13 tons of CO_2-eq per hectare in 2020.[15] In fact, he was the first farmer to be paid for sequestering soil carbon under the Australian government-regulated system.[16] Olsen's pasture cropping system is significant because if applied to permanent pastures and arable/croplands, it would sequester 63.8 gigatons of CO_2 annually (see appendix).

"No Kill, No Till"

Singing Frogs Farm, run by Elizabeth and Paul Kaiser, is a highly productive "no kill, no till," biodiverse, organic, agroecological horticulture farm on 3 acres (1.2 hectares) in Northern California. The key to their no-till system is to cover the planting beds with mulch and compost instead of plowing them or using herbicides, and planting directly into the compost, along with a high biodiversity of cash and cover crops that are continuously rotated to break weed, disease, and pest cycles.

According to Chico State University, the Kaisers have increased soil organic matter (SOM) by 400 percent—from 2.4 percent to 7–8 percent, with an average increase of about .75 percentage points per year—in six years. This farming system could apply to more than 80 percent of farmers worldwide, as most have fewer than 5 acres (2 hectares). If the increases in Singing Frog Farm's soil carbon were extrapolated globally across arable and permanent croplands, it would sequester 179 Gt of CO_2 per year.

Biologically Enhanced Agricultural Management (BEAM)

Biologically Enhanced Agricultural Management (BEAM), which was developed by Dr. David Johnson and Dr. Hui-Chun Su Johnson of New Mexico State University, produces compost with a high diversity of soil microorganisms. Over the years, I have gotten to know David and his wife, Su, and have seen the compost system in a range of climates in Mexico and the United States. The BEAM system aligns well with research showing how organic carbon-based inputs such as composts encourage higher proportions of root exudates than synthetic water-soluble chemical fertilizers.[17] Multiple crops grown with BEAM have achieved very high sequestration levels and yields. Research published by Johnson and colleagues shows an annual average capture and storage of 10.27 tons of soil organic carbon (SOC) per hectare annually.[18] These results are currently being replicated in other trials.

These figures mean that BEAM can sequester 37.7 metric tons of CO_2 per hectare per year. BEAM can be used in all soil-based food production systems, including for annual crops, permanent crops, and grazing systems, and it can work in arid and semiarid regions. If BEAM were extrapolated globally across agricultural lands, it would sequester 185 Gt of CO_2 annually.

How Does the Flores González Brothers' Agave System Fit In?

The primary significance of the Flores González brothers' and Dr. Juan Frias' development of the agave agroforestry system is that most of the species in other agroforestry systems could also be fermented to remove saponins, lectins, and other toxic compounds and used as high-quality feed, as has been demonstrated with agave. Selective harvesting for feed would increase the productivity of all of these systems. For example, the agave agroforestry system can be applied to Savory's adaptive multi-paddock grazing system to provide forage in the drier and cooler seasons when the pasture grasses do not grow and can be easily overgrazed.

In terms of climate, research by Dr. Mike Howard shows that the agave agroforestry system can sequester 8.7 metric tons of carbon dioxide per hectare per year.

This is without counting belowground SOM sequestration or the amount of carbon sequestered by the companion trees. Extrapolated globally across the 3.3 billion hectares (8.2 billion acres) of permanent pastures, the agave agroforestry system could sequester 28.7 Gt of CO_2 annually. This is slightly more than the 28 Gt of CO_2-eq per year that is currently emitted, and so implementation of this system would start to reverse climate change and regenerate the climate.[19] (See appendix for methodologies to calculate these figures.)

This possibility does not include the extra functions that such a system provides, such as cooling the region through regenerating forests and permanent pastures. The shading and rehydrating of the landscape will reduce the ambient temperature. The potential for soil carbon sequestration is very high due to the role of deep roots excreting around 30 percent of the carbon compounds created through photosynthesis into the soil. This could sequester another 2.6 tons of CO_2 per hectare.

Agriculture: From Significant Problem to Primary Solution

Critics of regenerative agriculture's potential to draw down CO_2 often argue that all soils have an optimum SOM level, which cannot be exceeded. Research has shown, however, that this is not a function of the soil. It is a function of the soil management system. The latest research shows no upper limit for SOM with the correct management.[20]

A substantial body of evidence, starting in 1904, shows how root exudates feed organic carbon compounds to the soil microbiome, thereby increasing SOM. The key is management systems that maximize photosynthesis to capture CO_2 and convert it into numerous organic compounds.[21]

Many long-term trials show that organic farming systems have higher rates of SOM increases than industrial/conventional systems.[22] Organic farms do not use synthetic chemical fertilizers that cause decreased SOM and produce lower percentages of organic carbon-based root exudates.

The evidence shows that agriculture needs to change from chemically intensive to biologically intensive. The new paradigm reduces and ultimately avoids the use of synthetic chemicals. Plant biology and living soil science must be at the forefront of developing these nature-based solutions.

A general rule is that the soil should be covered with the maximum amount of living plants for as long as possible in the growing season. Dead plants and bare soil do not photosynthesize, so the most productive regenerative systems avoid killing plants as weeds with herbicides and excessive tillage. Instead, plants are managed as ground covers and cover crops to build soil fertility by maximizing root exudates. Various strategies are used to manage weeds and turn them into cover crops to build fertility.[23]

Another significant issue with criticism of regenerative agriculture's capacity to draw down CO_2 is a focus on datasets derived from industrial-agricultural systems that decrease SOM or have meager increases. Such research papers correctly show that these systems are unsuitable for scaling up to achieve the sequestration levels needed to mitigate climate change. Their extensive datasets show that the current industrial/conventional agriculture systems are unsuitable for increasing SOM to the levels required to sequester greenhouse gases. These critics often cherry-pick studies that support their preconceived biases and omit datasets showing meaningful increases in SOM. Such critics have ignored an extensive body of published studies showing that systems under regenerative agriculture, such as organic agriculture and AMP grazing, can sequester significant amounts of CO_2 and increase SOM over many decades.[24]

Many such examples are derived from industrial farming systems that use synthetic nitrogen fertilizers, which long-term data show deplete SOM. For example, researchers analyzed the results of a 50-year agricultural trial and found that applying synthetic nitrogen fertilizer resulted in all the carbon residues from the crop disappearing and an average loss of around 10,000 kilograms of soil organic carbon (SOC) per hectare (10,000 pounds per acre). This equates to greenhouse emissions of 36,700 kilograms of CO_2 per hectare (36,700 pounds per acre) over and above the many thousands of kilograms of crop residue that are converted into CO_2 every year. Multiple researchers have found that the higher the application of synthetic nitrogen fertilizer, the greater the amount of SOC lost as CO_2.[25]

It's All About Soil

Soils are the most significant carbon sink after the oceans. We already have too much CO_2 in the oceans. According to Professor Rattan Lal of Ohio State University, over 2,700 Gt of carbon is stored in soils and 575 Gt in biomass (trees and forests) worldwide.[26] The soil holds more than twice the amount of carbon as the atmosphere (848 Gt) and biomass combined.

Regenerating ecosystems and soil carbon is the key to reversing climate change. Stopping deforestation and destructive forest fires and planting more trees will improve the environment and must be done. Combined with soil carbon sequestration, these actions can reverse emissions long before the 2050 net-zero goal.

The best systems maximize photosynthesis to increase the capture of CO_2 and store it in the soil as SOM through organic matter biomass and root exudations.[27] Depending on the management system and the species, root exudates can distribute 10 percent to 40 percent of the carbon from photosynthesis into the soil while the plants grow.[28] The carbon compounds from root exudates penetrate deeper into the soils due to the depths of the roots than those of aboveground or tilled biomass. Aboveground and tilled biomass can rapidly oxidize into CO_2. Systems with deeper roots are ideal as their exudates build more durable SOM because deep soil carbon tends to be more stable.[29]

The key is ensuring that agricultural systems have photosynthesizing plants for the most prolonged periods in their climates. This is achieved by using a diversity of correctly managed species to ensure they can capture the maximum amount of sunlight per acre as the energy needed to convert CO_2 into the organic molecules that build SOM through the soil microbiome. Permanent covers of living plants and limited tillage systems are the best way to increase SOM.[30]

"Back-of-the-Envelope" Calculations

A simple "back-of-the-envelope" calculation shows that transitioning a small proportion of agricultural production to best-practice regenerative organic systems will sequester enough CO_2 to draw down more than the current emissions and result in negative emissions. (See appendix for details and methodology on calculations.)

- Ten percent of arid and semiarid drylands under the agave agroforestry system could sequester 2.9 Gt of CO_2 per year.
- Five percent of global agricultural lands regenerated by the BEAM organic compost system could sequester 9.2 Gt of CO_2 per year.
- Five percent of smallholder farms across arable and permanent croplands using Singing Frogs Farm's bio-intensive organic "no kill, no-till" system could sequester 8.9 Gt of CO_2 per year.
- Ten percent of grasslands under regenerative grazing could sequester 9.9 Gt of CO_2 per year.
- Ten percent of agricultural lands using pasture cropping could sequester 6.4 Gt of CO_2 per year.

This would result in 37.3 Gt of CO_2 per year being sequestered. This is better than the 28 Gt of CO_2-eq per year that is currently emitted, and would therefore start to reverse climate change and regenerate the climate.

Combining these regenerative systems is not double- or triple-counting. Many permanent pastures are unsuitable for cropping and can only be used for grazing. Pasture cropping can be used in most arable and grazing systems where machinery can be safely operated and there is sufficient soil moisture in the rainy season to grow an annual crop. BEAM can be used in all systems. The different systems give flexibility and more options for adoption by landholders. Furthermore, a 10 percent adoption rate is a realistic goal, especially for working with early adopters.

The Potential to Increase Soil Organic Carbon Is Greater than We Think

In fact, increases in SOC are typically much higher than what is reflected in most published literature.[31] For this reason, some authors and researchers express skepticism about the credibility of reports. However, the material and methods used in the above examples are published and can be replicated. They are evidence-based systems. Dismissing them based on an opinion is the opposite of science. The only way to prove or disprove these results is to replicate the material and methods and see the results.

There is an urgent need to transform agriculture from a significant source of greenhouse gases, vapor pressure deficit, and local warming to a major mitigator. Agriculture needs to contribute to the suite of the solutions necessary to reverse emissions to avoid 3.6°F (2°C) or more of significant warming and all the associated problems.

The above examples of regenerative agricultural systems and other outliers have the most potential. They should be the focus of future research and replicated in different climates and soil types. These methods should be scaled up to sequester CO_2 if increasing results are positive. Further research should be prioritized to improve these systems.

The Many Benefits of Trees and Forests

Planting trees in urban and cleared areas can have a dramatic cooling effect. A review of 308 studies found that tree canopy shading significantly affected daytime heat. Forests were, on average, 2.9°F (1.6°C) cooler than non-forested areas. Grassy areas were an average of 1.1°F (0.6°C) cooler than non-vegetated areas. This is significant as the world is about to become 2.7°F (1.5°C) warmer than it was in 1750. Just planting trees to shade and cool all our towns, cities, and other regions would significantly reverse the increase in heat from the major greenhouse gases.

A review of the research on non-carbon effects and benefits found that deforestation has multiple critical impacts beyond the loss of the CO_2 drawdown function of forests. Healthy forests release a range of volatile organic compounds that have an overall cooling effect on our climate, mainly by blocking incoming solar energy. Destroying forests eliminates this cooling effect and adds to warming. The United Nations Convention to Combat Desertification (UNCCD) states that industrial agriculture is the cause of 80 percent of deforestation.[32]

According to author and journalist Fred Pearce, researchers have calculated the actual contribution of deforestation to global warming since 1850 to be as much as 40 percent and that the current rate of tropical deforestation could add 2.7°F (1.5°C) to global temperatures by 2100 even if there were zero fossil fuel emissions. Forests moderate local climates by keeping their local environments cool. They do this partly by shading the land and releasing moisture from their leaves. This process, called transpiration, extracts energy from the surrounding air, thus cooling it. A single tree can transpire hundreds of liters of water in a day. Every 100 liters (26 gallons) has a cooling effect equivalent to two domestic air conditioners daily.

Monitoring rapidly deforesting regions of the tropics has recently shown the effect of losing this arboreal air conditioning. Sumatra has been losing forests to palm oil cultivation faster than anywhere else. A study found that since 2000, surface temperatures have increased by 2°F (1.1°C) compared with 0.8°F (0.45°C) degrees in forested parts. Another study found temperature differences between forest and clear-cut land of up to 18°F (10°C) in parts of Sumatra. Research in the Amazon found a difference of 5.4°F (3°C) between the forested Xingu Indigenous Park and surrounding croplands and pastures.[33]

Researchers at the Carnegie Global Ecology Department concluded that increased evaporation from forests and lakes had an overall cooling effect on the global climate. Increased evaporation tends to cause clouds to form low in the atmosphere, reflecting the sun's warming rays out into space. This increased albedo effect is an example of climate forcing that lowers global and local temperatures. It shows that increasing forest cover, rather than decreasing it through clearing and forest fires, will cool the planet by changing its hydrology. The ability of forests to draw down CO_2 further assists in this.[34]

Researchers have also found that the reforestation of the eastern United States over the last century has had a cooling effect that has resulted in a lack of regional warming over that same period. This stands in contrast to warming trends across the rest of North America during the same period. The study in question shows that forests across much of the eastern United States have a substantial adaptive cooling benefit for air temperatures. Ground and satellite-based observations has showed that these forests cool the land surface by 1.8–3.6°F (1–2°C) annually compared to nearby grasslands and croplands, with the strongest cooling effect during midday in the growing season, when cooling amounts to 3.6–9°F (2–5°C).[35]

The regeneration of tree cover is one of the most effective strategies for climate change mitigation. Researchers mapped the global potential tree coverage and found that an extra 0.9 billion hectares (2.2 billion acres) of canopy cover could store 205 gigatons of carbon. This highlights that global tree regeneration is one of the most effective carbon drawdown solutions.[36]

The destruction of valuable forest ecosystems is a significant climate forcing that contributes to global warming, whereas regenerating these forests would cool the planet.

Drawdown

CO_2 emissions have increased by about 2.87 ppm annually since the Paris Agreement of 2015. This means that 28 Gt of CO_2-eq per year needs to be sequestered from the atmosphere annually to stabilize the current level. (See appendix for the methodology.) We have to draw down more than this and then reverse emissions to reduce the levels of CO_2 to keep the temperature rise to 2.7°F (1.5°C). Net zero by 2050 will see the world heat up to more than 3.6°F (2°C) as the world has already passed the threshold of 450 ppm CO_2-eq.[37]

We don't have to wait for 100 percent of agriculture to change. That is unrealistic. Just a percentage of early adopters applying best-practice regenerative systems to their landholdings can significantly contribute to reversing emissions needed to regenerate ecosystems and stabilize the climate.

Actual amounts of carbon sequestered will vary because best practice systems rarely scale up evenly. That said, we do know that scaling up these regenerative systems can make an enormous difference to both the global and the local climate, not only by sequestering CO_2, but by cooling local ecosystems and restoring shade and hydrology.

The regenerative agriculture systems profiled in this chapter are only a few among many. In fact, many emerging systems, mainly perennial agroforestry systems, have the potential to achieve even higher increases in SOC. Even if the results were half that of my calculations, the outcome would be impressive and a massive contribution to drawing down carbon dioxide and stabilizing the climate. These evidence-based systems can make an immense difference in the lives of thousands of people. In the following chapter, we'll look at a few real-world examples of these systems in practice—including the agave agroforestry system—and the impact they've had on climate, local ecosystems, the local economy, and the lives of everyday people.

CHAPTER FOUR

Success Stories

The power of regeneration is impressive. In April 2012, my wife, Julia, and I flew from our home in Australia to northern Ethiopia to visit the Tigray Project. My friend Hailu Araya Tedla from the Institute of Sustainable Development, who worked on the project, drove us there from our hotel in Aksum, the ancient city of impressive ruins that Ethiopians believe was the home of the Queen of Sheba.

I could see evidence of the Tigray Project miles before we reached the community. The lush, green revegetated sections of mountain ranges stood out in stark contrast to the rest of the barren, degenerated landscape. This desert is a man-made ecological and sociological catastrophe, like most deserts in Africa, the Middle East, and Asia. Such landscapes were once fertile rangelands that supported vast herds of animals and a rich diversity of trees, grasses, herbs, flowers, and other species. Their abundance allowed the rise of great civilizations across the region, such as Assyria, Ethiopia, Egypt, Carthage, Persia, the Indus Valley civilization, Greece, and Rome. Unsustainable agricultural practices of inappropriate tillage, continuous monocultures, non-rotational overgrazing, and soil loss increased the vapor pressure deficit (VPD), drying out the regions. This turned large sections of these fertile areas into deserts and barren semiarid farmlands, causing crop failures and resulting in famines and the collapse of these great empires. This is man-made climate change.

North Africa was once the breadbasket of the Roman Empire. The Roman Warm Period that started around 200 BCE was a time of good, reliable rain, resulting in bumper crops and prosperity. This gave Rome the food and income to employ hundreds of thousands of soldiers, conquering the neighboring countries to become the greatest of the Mediterranean empires.

Around 150 CE, the climate started to dry due to overgrazing and clearing land for cropping, resulting in an increase in the VPD. The fertile fields of North Africa dried out and were eroded away by wind and rain, turning them into deserts, causing crop failures. The empire could no longer feed its armies on the outer fringes and was forced to withdraw them. The continuing expansion of degenerated farming lands further reduced crop yields, forcing the empire to continuously reduce the size of its armies.

As the empire contracted, border conflicts and civil wars increased. Rome was conquered, but not by another great empire; it was too weak to fight off one of its small marauding armies in a civil war, whose leader, Constantine, moved the capital to Constantinople, present-day Istanbul, in Turkey, in 330 CE. Rome was no longer the center of a great empire covering most of western Europe and the Mediterranean, and the once fertile fields of North Africa still remain arid fields and deserts.

In Central America, the great Mayan cities such as Palenque, Tikal, and Copán in the southern lowlands of the Yucatán Peninsula likewise declined over a period of about 200 years from 800 CE. Researchers have determined that there was a sustained decline in annual rainfall—between 41 and 54 percent—during this period.[1] Richardson Gill's *The Great Maya Droughts: Water, Life, and Death* describes how the collapse of the southern Yucatán Maya civilization was driven by an unrelenting drought that killed millions of Maya people and initiated the abandonment of the great cities, towns, and farms of the southern Yucatán.[2]

There are many theories about what caused the drought, such as increases in solar radiation, solar cycles, and volcanos. These theories do not explain why the northern Yucatán and the mountain Mayan civilizations didn't collapse. In fact, they were thriving until the Spanish colonial invasion decimated them. Since the drought was localized to the southern Yucatán lowlands, widespread global influences were not the main cause. The most logical explanation is that clearing the forests plus the loss of soil organic matter and bare soils heated up the land and caused a prolonged period of increased VPD. The climate restabilized after the cities and farms were abandoned, and the forests regenerated.

This climate catastrophe is repeating now, however, with the clearing of huge swathes of the Yucatán and mountain rainforests in Mexico and Guatemala. The region is suffering from intense and prolonged droughts again. The crop failures result in farmers abandoning their farms, forcing them to migrate north to the United States.

In southeast Asia, the Khmer civilization was established in 802 CE. Its capital city, Angkor, was on the Tonlé Sap Lake in present-day Cambodia. The Khmer Empire ruled over much of mainland Southeast Asia and was at the height of its power and wealth by the beginning of 1100 CE, when it started building the main Angkor Wat temple.

Archeologists using radiocarbon dating found that the population then started declining, with abandonment in some areas from late 1100 CE to early 1400 CE. They have established that there was a drought from 1345 to 1374, causing a decrease in the population and a decline in the maintenance of the moats in the 1300s. The population started to increase around Angkor in the late 1300s until another drought that occurred between 1401 and 1425.[3]

Some researchers assume that failed monsoons caused the droughts. The monsoon affects the whole south and southeast Asian region from India to southern China and south through Indonesia. The small Thai kingdom of Ayutthaya, bordering the Khmer Empire on the northwest in present-day Thailand, was unaffected by the droughts that reduced the Khmer population, showing that the monsoon had not failed across Southeast Asia. Ayutthaya continued to grow and thrive. The Thai Kings of Ayutthaya, having abundant crops to feed and employ armies, expanded their kingdom by beating the Khmers in battles. They captured and occupied Angkor in 1431. This was the end of the Khmer Empire. As the forests regenerated and restored the climate, people returned slowly to Angkor, now present-day Cambodia.

It is clear that failed monsoons didn't cause the droughts that impacted the Khmer Empire. The droughts were local events caused by the clearing of the forests, the loss of soil organic matter, and bare soil heating up the land and causing the increased VPD that dried out the region.

The Khmers, like the Romans, Maya, and many other civilizations, were weakened by food shortages due to drought. These man-made climate disasters were not caused by a surplus of CO_2 and other greenhouse gases. The bare soil from their agricultural systems caused the increased VPD that dried out the regions, creating droughts and deserts. The climate was restored in the regions where the forests were allowed to regenerate.

Agave and mesquite growing wild in the Mexican high desert. *Photo courtesy of André Leu.*

José (black jacket) and Gilberto Flores González (white shirt). *Photo courtesy of Morgan Weyrens Welch.*

Ronnie Cummins in front of mature agaves. The lower leaves can be sustainably harvested to make silage. *Photo courtesy of Vía Orgánica.*

Dr. Juan Frias and André Leu in front of mature agave plants at the Vía Orgánica Ranch. *Photo courtesy of André Leu.*

Feeding agave leaves into the shredder. *Photo courtesy of Vía Orgánica.*

Dr. Juan Frias training farmers in how to make silage from harvested agave leaves. *Photo courtesy of Vía Orgánica.*

Ronnie Cummins next to sealed containers of silage. *Photo courtesy of André Leu.*

Goats eating agave silage. *Photo courtesy of André Leu.*

Chickens eating agave silage. *Photo courtesy of Vía Orgánica.*

The agave agroforestry system about one year after planting. *Photo courtesy of André Leu.*

At the Africa Centre for Holistic Management in Zimbabwe, a herd of cattle moves through the vegetation by the careful guidance of their herders. *Photo courtesy of Bobby Gill, Savory Institute.*

A fence-line comparison near the Karoo Desert in South Africa. At left, a highly vegetated landscape thanks to properly managed livestock. At right, a desertified landscape shows the effects of overgrazing. *Photo courtesy of Norman Kroon.*

Farmers learn how to propagate agaves at the Vía Orgánica ranch. *Photo courtesy of Vía Orgánica.*

A farmer stands in his rain-fed chili, pumpkin, and sorghum field. The previously bare hills in the background have been regenerated with tree covers, and the previously intermittent creeks now flow permanently. *Photo courtesy of André Leu.*

An orchard of rain-fed mangos that also supports cattle grazing. Before the region was regenerated, it was too dry to grow mangos. *Photo courtesy of André Leu.*

Mayor Rommel C. Arnado's organic farm has been established as an education center to train farmers in best-practice tropical organic systems. *Photo courtesy of André Leu.*

Dr. Vandana Shiva talking to an organic cotton farmer. The healthy organic cotton crop was grown from the native Indian seeds she saved. *Photo courtesy of André Leu.*

Spinning organic cotton. *Photo courtesy of André Leu.*

Hand-printing patterns on the cloth with natural dyes. *Photo courtesy of André Leu.*

The finished organic cotton saris are sold to provide all the farmers and co-op workers with a decent living wage. *Photo courtesy of André Leu.*

The eroded Loess plateau landscape before it was regenerated. *Photo courtesy of 黄河山曲 / Wikimedia Commons.*

The provincial and national governments have funded the establishment of organic apple orchards as an environmentally friendly farming system that provides farmers with a high income. *Photo courtesy of André Leu.*

An organic apple-farming village. The regeneration of the region, combined with premium income, has lifted some of the most impoverished farmers out of poverty. *Photo courtesy of André Leu.*

Regenerating ecosystems, including agroecosystems, can reverse man-made climate change. The good news is that there are many excellent examples and multiple ways to do it.

The Tigray Project, Ethiopia

The Tigray Project in Ethiopia has proven that highly degenerated regions can be regenerated in a few years, ending undernourishment, poverty, and environmental degradation. It is simple, cheap, and does not require new technology. We can do it now with what we already know. We need the funds to be redirected from industrial degeneration to regenerative nature-based solutions.

Ethiopia was once one of the most prosperous countries in Africa. Its civilization goes back thousands of years alongside Ancient Egypt, India, and China, with impressive cities and ancient monuments. It has a unique writing system, music, art, dance, cuisine, and branches of Christianity and Judaism. Much of the country was once highly fertile; in fact, it is one of the places where agriculture began through the domestication of numerous indigenous crops such as coffee, teff, various millets, sorghum, and the banana species *Ensete ventricosum*. Early reports of Ethiopia recorded a land of plenty and abundance in food and tradable items such as myrrh, gold, ivory, and precious gems.

This all changed during colonialism when the fascist dictator, Benito Mussolini, invaded the country in the 1930s, enslaving the people, looting resources (including food), and decimating farming systems. Since then, a series of corrupt and inept governments, combined with drought and civil wars, have presided over massive famines that have killed thousands. Bob Geldof, the singer of the Boomtown Rats, was so moved by the images of famine in Ethiopia that he got together with Midge

Ure from Ultravox and formed the charity "Band Aid." They wrote the song "Do They Know It's Christmas?," which was recorded by a who's who of the British music industry in 1984 and became a global number-one hit. All the royalties were used to fund projects in Ethiopia to help the victims of famine.

This was not enough, so Bob and Midge organized the Live Aid benefit concert held at Wembley Stadium in London and JFK Stadium in Philadelphia on July 13, 1985. These concerts featured the greatest names in the UK and the US rock industries. Around 1.5 million people watched it on TV, and it raised millions of dollars to feed people in Ethiopia. Unfortunately, this was a temporary solution. People continued to starve due to corruption, incompetence, conflicts, and droughts.

A new government was elected in 1995. Officials of this new government asked Dr. Tewolde Egziabher and Dr. Sue Edwards of the Institute for Sustainable Development (ISD) to design a project to assist poor farmers in marginal areas in improving the productivity of their land and rehabilitating the environment. Dr. Egziabher and Dr. Edwards then started The Tigray Project in 1996 in four local communities, with funding from Third World Network. Later, with funding from several donor agencies, the Ethiopian Bureau of Agriculture and Rural Development assisted in scaling up the project's scope so that more regions in Ethiopia could adopt the practices. Dr. Egziabher and Dr. Edwards, both now deceased, became good friends of mine. I visited their home several times in Addis Ababa, and participated in various international events with them to promote regenerating landscapes, biodiversity, and farming to end poverty and restore the climate.

At the time the project was initiated, Tigray was an area regularly affected by famines that killed many people. ISD cooperated with local farmers to revegetate their landscape to regenerate the local ecology and hydrology. The farmers did not stop grazing as their animals were an essential source of food and income. But Dr. Egziabher and Dr. Edwards worked with the farmers to develop rotational grazing plans that allowed time for vegetation to recover and grow.

Before long, the denuded hills became green with trees, shrubs, and grasses. This restored the hydrology, so the rivers and creeks began to flow permanently instead of seasonally. The vegetation cooled and moderated temperatures in the local area, reducing the extremes of both heat and cold.

Revegetating marginal areas such as water courses, gullies, steep slopes, roadsides, laneways, and field borders, and sustainably harvesting the biomass, provided a steady source of nutrients in addition to those generated through good organic practices in the fields. Revegetation of marginal areas is particularly important for building up soil fertility, and—combined with growing functional biodiverse species, such as deep-rooted legumes for nitrogen production, host plants for the natural enemies of pest species, and taller species as windbreaks—these revegetated marginal areas provide a range of useful ecosystem services.

The biomass from this revegetation was sustainably harvested to make compost and feed biogas digesters. Biogas enabled energy independence in the villages by supplying all the gas needed for cooking and lights and reducing the need to cut down vegetation for cooking fires. The residues from the biogas digesters were composted and applied to the crop fields. The result after a few years was more than 100 percent increases in yields and better water use efficiency.

The farmers used the seeds of their landraces, developed over millennia and locally adapted to the climate, soils, and significant pests and diseases. The best of these farmer-bred varieties proved very responsive to producing high yields under regenerative conditions.

A significant advantage of this system was that the seeds and the compost were sourced locally at no or little cost to the farmers. In contrast, the industrial agricultural systems' seeds and synthetic chemical inputs had to be purchased. Not only did the regenerative system have higher yields, but it also produced a better net return for the farmers. According to Dr. Edwards, the farmers who used compost earned $2,925 per hectare compared to $1,725 per hectare for farmers who used chemical fertilizers.

Using simple, appropriate regenerative farming methods, this project transformed a region that had been regularly affected by severe famines into a place of food surplus and relative prosperity. The people could now afford to eat well, buy clothes, send their children to school, pay for medical treatment, afford transport into town, and build adequate houses.

When Julia and I visited the Tigray Project, Hailu showed me rivers and creeks that were now permanently flowing, whereas previously they dried up for most of the year. The increased prosperity meant that young people were returning because they could secure good employment and build a future for themselves. For example, a group of returned youths began keeping bee hives in the revegetated hillsides, making a good living selling the honey.

Sue also told me that the women could buy new clothes. When I asked why this was important, she explained that their clothes were so old they had large holes, exposing their skin. The women felt indecent and would not socialize. Once they had new clothes, they began socializing and organizing community events. This regenerated the community, as well, and provided a considerable sense of well-being. The project transformed a region devastated by environmental degradation, droughts, lethal famines, and extreme poverty to one where the landscape, forests, and farms were restored, providing food security, sovereignty, prosperity, and the empowerment of well-being.

Sadly, as I write these words in 2023, Ethiopia has been thrown into another civil war. The central government, aided by the government of Eritrea, invaded Tigray, bombing the towns and cities, burning the fields, and massacring the people. It is estimated that hundreds of thousands of Tigrayan civilians have been killed. This ongoing tragedy is being ignored in the mainstream media. Regenerating peace is essential to the well-being of all societies and needs to be a priority. Wars and conflict cause more pain, suffering, and hunger than droughts and floods. They must be prevented. Community leadership can make a huge difference in building peace and ending conflicts.

From Arms to Farms, the Philippines

The municipality of Kauswagan on the island of Mindanao in the Philippines was once a desolate, war-torn region whose residents suffered from extreme poverty, violence, and hunger. In 1969, the Moro Islamic Liberation Front (MILF) formed following the Jabidah massacre, during which Philippine Army troops massacred a unit of Muslim soldiers. MILF initiated a series of campaigns, including terrorist and guerrilla warfare, to form an independent Muslim nation on the island of Mindanao. Several attempts to negotiate a peaceful settlement were undermined by influential members of the central government and judiciary, and by August 2008, the conflict escalated, with the Armed Forces of the Philippines deploying ten battalions comprising 6,000 soldiers. This phase of the war displaced over 600,000 people, and hundreds died.

Rommel Arnado was born and raised in Kauswagan, left for many years to work as a pilot in the United States, and returned to his birthplace in 2008. But back in Kauswagan,

Arnado was dismayed by the violence, the desperate conditions, the hundreds of thousands of displaced families, the hunger, and extreme poverty. He decided he must help improve the people's lives in his community. He was elected mayor in 2010 and immediately initiated a process of bringing people together. The community developed "The Sustainable and Integrated Kauswagan Area Development and Peace Agenda" (SIKAD PA), in which they identified and targeted the poorest among the poor communities and created opportunities for their voices to be heard. This humble beginning paved the way for multiple improvements in Kauswagan's residents' quality of life.

When I visited in 2023, Mayor Arnado told me the conflict's leading cause was extreme poverty and hunger. Feeding the people with healthy food and having a surplus to sell to bring in sufficient income to end poverty were the keys to solving the problems. Consequently, SIKAD PA gave birth to the "From Arms to Farms" program. This program developed organic farms based on regenerative and agroecological practices.

Mayor Arnado, a Christian, contacted his Muslim neighbors. They informed him how they needed irrigation for their crops, so he arranged for this to be supplied by various appropriate systems. He learned that they needed to improve soil fertility and have better pest control to achieve yields that could end hunger and poverty. This was when he realized that expensive synthetic chemical fertilizers and pesticides would put them further in debt and create a permanent poverty trap.

Organic agriculture was the solution so that soil fertility and pest management could be developed on farms for no or minimal costs. Organic experts were brought to Kauswagan to assist with the transition. A range of training and infrastructure programs were developed and scaled up. Partnerships of key players, including community leaders, municipal employees, officials, and townsfolk, worked with the Muslim farming communities.

Instead of conflict, Muslims and Christians cooperated to achieve community food security and prosperity. This started to build trust and respect between the warring communities.

Mayor Arnado approached the MILF guerrilla commanders about collaborating on a pathway to peace. He said, "I'm not asking you to give me your arms; I'm asking you to give me your hearts."

This started the process of the soldiers laying down their arms and developing successful organic farms. By 2016, the program had assisted over 600 former MILF soldiers and their families in developing diverse and productive organic farms. The rate of poverty in the area decreased to 40 percent.

According to Mayor Arnado: "The powerful transformation of our communities is shown in the dramatic drop [in the] poverty incidence rate from 69.6 percent in 2009 to 9.1 percent in 2019."

The children could attend schools, and all the communities in the municipality now had health services.

Influential people in politics and the judiciary had successfully undermined all the previous attempts to end the civil war. They tried to remove Mayor Arnado from office to end the peace program. Due to his immense popularity, Mayor Arnado's political opponents couldn't beat him in the ballot box, so they used a range of dirty legal tricks through the courts to try and get him jailed, and then later had the courts dismiss him from office for having a US passport, even though he renounced his US citizenship in order to be elected mayor. He fought and eventually won all these political battles.

A large proportion of Mayor Arnado's success was the widespread trust, respect, and love he had built with the people of Kauswagan. In only a decade, he had helped them transform their community from a wasteland devastated by violent warfare, which had displaced hundreds of thousands of people and resulted in hundreds of deaths, poverty, and hunger, to a place of peace, an abundance of food, and prosperity for most people.

When I was there, I was particularly moved by the speeches of former MILF commanders praising Mayor Arnado for ending the war by scaling up organic agriculture. These commanders and the men who fought under them now enjoy peaceful lives with their families on highly productive farms.

The farms I visited are some of the best I have seen anywhere. They are highly diverse agroforestry systems with numerous vegetables and herbs growing under canopies of coconuts and a diverse range of tropical fruits. Some farms cultivate rice, and most integrate a range of livestock including poultry, goats, water buffalo, and cattle.

These farms prove that the best way to nourish people—and bring about stability and peace—is through highly diverse regenerative organic farms, not industrial monocultures.

Navdanya, India

Most years, I visit Vandana Shiva's farm and headquarters, Navdanya, a picturesque oasis of biodiversity nestled in the Himalayan foothills near Dehra Dun in India. I teach several subjects at her Earth University course, the "A to Z of Agroecology," a life-changing course for many participants. I have also traveled around several regions of India with Vandana and seen some of her projects firsthand, the immense respect she engenders from the local communities, and the multiple benefits of these projects.

Vandana graduated with a PhD in quantum physics, but instead of pursuing a career as a scientist, her life was changed by the Chipko movement in her home region in the Doon Valley. The Chipkos were the original tree huggers of the 1970s, embracing trees to stop their destruction by loggers. Male and female activists both played critical roles in the movement; however, it was the mass participation of female villagers that dominated the nonviolent actions. They embraced the trees to stop the chainsaws.

Vandana, the daughter of a forester who spent much of her childhood in the forests of the Himalayas and other regions of India, could see the immense destruction occurring in her lifetime with the loss of thousands of acres of these irreplaceable, richly biodiverse ecosystems. She understood how the loss of these ecosystems was adversely affecting the local communities through landslides, floods, droughts, polluted waterways, dust, heat, and haze. She readily identified with their struggle and use of Gandhian nonviolence to save the trees and forests, and joined the movement as an active participant.

This started her path of Satyagraha, nonviolent noncooperation, to stop degenerative practices, corruption, exploitation, and the destruction of biodiversity and farming systems. She became an influential activist supporting, empowering, and regenerating local communities. She says, "Regenerative agriculture provides answers to the soil crisis, the food crisis, the climate crisis, and the crisis of democracy."

Her book, *The Violence of the Green Revolution*, has had a significant impact, showing that the narrative that the Green Revolution ended starvation and fed the world was a dangerous lie. The book opened the global debate about introducing industrial agricultural systems in developing countries. Vandana clearly articulated how introducing agriculture based on toxic synthetic pesticides and fertilizers and patented hybrid and GMO seeds was actually increasing poverty, debt, people being forced off their farms, and massive disruption of families and communities. Rather than solving world hunger, the number of food-insecure people increased yearly.

I have spent more than 50 years visiting thousands of farms worldwide and have seen the dramatic rise in the use of toxic synthetic chemicals, along with the increase in health problems and the impoverishment of rural communities due to debt and high-priced inputs. In many cases, the payments farmers receive for any extra yield rarely cover the money they have to borrow to purchase these expensive inputs, sending them into a perpetual spiral of debt.

Most farmers in the developing world do not use safety equipment like respirators or protective clothing and gloves. In some cases, they mix the pesticide with water, stir it with their bare hands, and then splash it over the plants from a bucket with bare hands because they cannot afford spray equipment or protective clothing. Often, farmers do not understand that the chemicals are toxic to humans. In India, the word for pesticide is *dava*, which means "medicine" in Hindi. Because synthetic pesticides are marketed and sold as "medicine for plants," many of India's 600 million farmers think these products are healthy as opposed to highly dangerous. Farmers often store the chemicals and mix them up in their houses, surrounded by family members and their food. The highest pesticide poisoning rates are among farmers, their families, farmworkers, and rural communities in the developing world.

Vandana coauthored *Poisons in Our Food* with her sister, Dr. Mira Shiva, a medical doctor, and their colleague, Dr. Vaibhav Singh. Their book gives numerous examples of how the misuse of pesticides in India has caused an epidemic of serious diseases. The book synthesizes the research on the link between disease epidemics like cancer and severe birth defects with the use of pesticides in agriculture in India. The most chilling example is Punjab, where the Green Revolution started in India. It is the center of a cancer epidemic, killing thousands of people. The number of people requiring cancer treatment has increased so much that the infamous "Cancer Train" is dedicated to taking thousands of people to hospitals for treatment.[4]

As an activist, Vandana not only talks the talk, she walks the walk. She is the prolific author of over 30 books on numerous subjects, including GMOs, climate change, exploitation by billionaire capitalists, toxic pesticides, regenerative organic agriculture, agroecology, and ecofeminism. One subject in which she has made an indelible mark is the importance of farmer-bred seeds. Vandana quickly realized that seeds were the key to annual crop production in agriculture. Without seeds, much of agriculture as we know it would not exist. She set up Navdanya as a project to save the seeds that farmers have developed over millennia. These seeds, known as "farmer landraces," are highly diverse and unique. Farmers have selected them because they are the best varieties for their soils and local microclimates. Research has shown that these seeds do far better than patented hybrid seeds under organic and agroecological systems.

Farmers have traditionally saved their best-performing seeds. This is how they continually improve their varieties. Also, saving their seeds meant they did not have to spend money buying them. The loss of these crucial varieties is the most significant extinction event on the planet, with thousands of unique and diverse landraces disappearing every year as governments bring in laws to make it illegal for farmers to save, swap, or trade their seeds. This drives farmers into debt by forcing them to purchase pricey patented hybrids and GMO seeds from agribusiness monopolies.

Modern hybrids only achieve high yields when they are treated with a lot of expensive inputs. These seeds need synthetic fertilizers, pesticides, and irrigation to achieve maximum yields. Without these expensive inputs, their yields decline substantially. Since farmers in the Global South can rarely afford to pay for all the expensive inputs, they use less. With fewer inputs, they don't achieve the expected yields and are driven further into debt. Additionally, most of these new varieties are not resilient to extreme weather events. Consequently, crop failures are common. These farmers often leave their farms to migrate for work to avoid starvation.

Vandana set up Navdanya more than 30 years ago. She chose the name *Navdanya* because it means "nine seeds" in Hindi, symbolizing the protection of biological and cultural diversity. It also means the "new gift" in the context of seeds as commons, based on the right to save and share them.

Navdanya has set up 150 community seed banks in 22 states of India. It has run training programs for about 750,000 farmers in seed sovereignty, food sovereignty, and regenerative organic agriculture. It helped establish India's most extensive direct marketing and fair-trade organic network. On top of that, Navdanya has collected, saved, and conserved thousands of unique farmer-bred seed varieties (farmer landraces), including 4,000 rice varieties and forgotten food crops such as millets, pseudo-cereals, and pulses.

These farmer landraces have unique characteristics—they include, for instance, salt-resistant rice varieties. Large areas of the coast of Tamil Nadu were inundated with seawater due to the Indian Ocean tsunami on December 26, 2004. Over 200,000 people were killed in Indonesia, Thailand, Myanmar, India, and Sri Lanka by this tragic event. The surge of waves hitting Tamil Nadu is estimated to have killed over 10,000 people. The receding seawater left the coastal rice paddies contaminated with salt, severely damaging the crops. Navdanya's saved salt-tolerant rice varieties were sent to the affected communities. This enabled them to grow rice again in the salt-affected fields to feed their communities and earn much-needed income.

India is one of the source cultures of agriculture, independently domesticating various crops and livestock around the same time as numerous crops were being domesticated in China, Southeast Asia, Papua New Guinea, Australia, Iran, Turkey, Iraq, Egypt, Ethiopia, Uganda, Zimbabwe, and throughout South, East, West and North Africa, Mexico, Peru, Brazil, and numerous other regions in the Americas—over 10,000 years ago.

One of the crops was cotton. The cotton species in India differ from the current commercial species domesticated in the Americas. During the Green Revolution, the Indian government promoted hybrids developed from American species, and these started to replace the traditional Indian varieties. The hybrids were susceptible to attacks from insects and diseases and required large amounts of toxic pesticides constantly sprayed on them to get yields. The pests and diseases rapidly became resistant to all the chemicals, and crops failed despite massive increases in toxic pesticide use. Genetically modified cotton was promoted because it was supposed to be insect-resistant and didn't need to be sprayed.

Pests and diseases soon became resistant to the GMO cotton, as well, causing widespread crop failures. The high cost of the GMO seeds and the other necessary inputs meant the farmers not only lost income from the crop failures but were now in substantial debt to loan sharks. When farmers are faced with the threat of losing their farms, social problems, such as substance abuse, depression, domestic violence, and suicide, multiply.

Vandana has written and spoken extensively on the epidemic of farmer suicides after crop failures in India. She has been criticized by some in the media, government, and academia about this, saying that she is inventing a nonexistent problem. However, from my experience listening to farmers across India, especially the wives and children who have lost husbands and fathers, and visiting the affected regions and farms, I know there is a genuine suicide epidemic. This is a massive social problem that governments are ignoring.

The late Dr. Mae Wan Ho from the Institute of Science and Society researched Indian government records, which showed that 182,936 farmers committed suicide in India between 1997 and 2007. Nearly two-thirds occurred in five states—Maharashtra, Karnataka, Andhra Pradesh, Madhya Pradesh, and Chhattisgar—with one-third of the country's population. Around 8 million people left farming between 1991 and 2001. The rate of suicides and farmers forced to abandon their farms continues to rise.[5]

The GMO cotton crops pollinated with the Indian varieties, contaminating them with genetically modified DNA. Fortunately, Navdanya had saved uncontaminated farmer-bred Indian cotton varieties. Vandana started a program of growing organic cotton from these seeds. This ensured that the organic cotton was not contaminated with pollen from the GMO cotton. The project increased the number of seeds and farmer families growing organic cotton. The yields and fiber quality were better than that of the GMO cotton. After being trained in good practice organic agriculture using manures and composts to build soil health and fertility, the farmers did not have to pay enormous prices for patented seeds or synthetic pesticides and fertilizers. Organic cotton did not suffer pest and disease damage. Higher yields and lower production costs for organic cotton meant better profits at the end of the season, bringing well-being to the farmers and their communities. Not one organic cotton farmer has committed suicide.

When Ronnie Cummins and I visited these organic fields with Vandana and compared them to the nearby GMO fields, it was clear that organic cotton was superior. The plants were more than twice the size of the GMO plants and had significantly more cotton bolls. The organic fibers were longer, more robust, and superior for ginning and spinning. There were no pests or diseases in the organic fields.

Navdanya started organic cotton supply chains, helping communities gin, spin, weave, and make garments. One of the examples Ronnie and I visited was a women's co-op at Wardha in Gangasagar Ward in Maharashtra State, in the center of India. This co-op was established with Navdanya's assistance to help women and their families who had lost their husbands and/or fathers through suicide. It is a complete supply chain where the organic cotton is collected from the farms, cleaned, and ginned. The cotton is then spun and woven into fabrics. These fabrics are dyed, pattern-printed, and colored in various styles. They are cut and tailored in saris, pants, shirts, scarves, and many other types of clothing.

The co-op has a shop where the finished products are sold to the local community. The income is distributed back to the employees to ensure everyone receives a fair wage for their labor. It is an excellent example of a holistic system that can be developed to regenerate agriculture, the environment, and communities by empowering people and ensuring prosperity and well-being for all.

This should be contrasted with the fast fashion industry that uses toxic herbicide-sprayed GMO cotton mixed with toxic synthetic plastic fibers in sweatshops that degenerate the environment, cause serious ill health, and exploit all the workers in the system.

Among the numerous books Vandana has authored, *Health Per Acre*, written with Dr. Vaibhav Singh, is a book that should be on everybody's reading list. This book compares the nutrition produced per acre in India's highly biodiverse agroecological, regenerative, organic farms (with multiple food crops) against the typical industrial farms that produce a few commodity crops. This is an important comparison between the two systems that has never been done before. Most studies compare the yield of organic systems with the yield of the conventional system per acre as a monoculture crop, whereas in reality, the two production systems are very different. Most smallholder organic farms, especially in developing countries, use agroecological systems of multiple food crops, such as numerous vegetables, fruits, grains, pulses, and animal products like eggs, milk, and meat, to supply the local market. On the other hand, most industrial farming systems tend to produce monocultures that are primarily single-variety cash crops destined for distant commodity trading markets.

The research reported in *Health Per Acre* showed that when all the nutrients in the crops were analyzed—rather than the standard measurement of pounds of gross produce per acre (or kilos per hectare)—scientists found that the diverse organic system produced significantly more key vitamins, minerals, and other vital nutrients that are needed for a healthy diet as compared to the industrial system. Measuring total nutrition from a farming system in this way is a critical methodology needed to measure food security and nutrition rather than simply measuring calories. Most of these calories lack natural nutrition—the vitamins, minerals, and other nutrients we need for good health.

This is a genuine problem. Over 850 million people are classified as food insecure globally. This means there are periods of the year when they do not have any food to eat. More than 1 billion additional people consume enough calories but are nevertheless deficient in critical nutrients. For instance, hundreds of millions of women in rural India have anemia due to iron deficiencies in their diet. This leads to a whole suite of health and reproductive problems. The reality is that iron deficiencies in rural women can easily be corrected by growing a few green leafy vegetables and including them in their daily diet. It is the same with beta-carotene deficiencies. Growing leafy and yellow/orange fruits and vegetables can quickly correct this. The research published in *Health Per Acre* showed that converting all of India's farms into biodiverse regenerative organic farms would correct all these deficiencies and provide enough nutrition to feed two Indias with a diverse diet of healthy, nutritious food.[6]

Dr. Vandana Shiva shows that transitioning agriculture to regenerative and organic systems using the science and practices of agroecology can provide good nutrition, stop the immense harm caused by toxic synthetic pesticides and fertilizers, and bring prosperity and well-being to farmers and their communities.

The Loess Plateau, China

John D. Liu is an ecologist, filmmaker, and one of the world's most eloquent spokespersons for ecosystem regeneration. Based in Beijing, China, John has worked with the BBC, National Geographic, Discovery, PBS, and other networks. He has written, produced, and directed films on grasslands, deserts, wetlands, oceans, rivers, urban development, the atmosphere, forests, endangered animals, and poverty reduction.

John's most significant film documented the rehabilitation of the Loess Plateau. The Loess Plateau Watershed Rehabilitation Project started in 1995 over 247,000 square miles (640,000 square kilometers) of the upper and middle reaches of China's Yellow River, named after the eroded yellow-brown soil that muddies its water.

Agriculture arose in the region over 10,000 years ago and led to the rise of the Han, Qin, Tang, and other Chinese dynasties. The forests and rangelands were cleared to plant crops, and the grasses and bushes were overgrazed by domesticated livestock. The fragile soil was eroded by wind and rain, resulting in the loss of the topsoil that held the soil organic matter. The loss of organic matter reduced the ability of the soil to hold moisture and damaged the structure of the soil. This resulted in erosion, forming massive gullies as the water ran off with the soil instead of being absorbed. This silted up the river, raising its bed so that it regularly flooded. The drier, barer soils radiated heat, creating VPDs over the whole region, causing huge dust storms, regular droughts, and famines. The area became an impoverished, man-made desert.

John documented the process of regenerating the Loess Plateau. The central government passed a national law banning unauthorized tree-cutting. Planting crops on steep slopes and free-range livestock grazing were banned to stop erosion and the denuding of vegetation. Livestock were required to be managed in ways that ensured the ecosystem could regenerate.

A critical part of the project involved working in partnership with the local people and educating them on the multiple benefits of regenerating the region. Many were employed in the rehabilitation efforts, and thus became active partners. Ensuring that local farmers had land tenure was important, along with clearly outlining their rights and responsibilities. The better land was used for agriculture, and the marginal land was planted with drought-tolerant native species to restore the natural ecosystem. Dams and other water-storage systems were developed to supply both the farms and the regenerating ecosystems with water to ensure that there was a continuous vegetation cover of grasses, shrubs, and trees to restore hydrology and prevent erosion.[7]

According to the World Bank, more than 2.5 million people in four of China's poorest provinces—Shanxi, Shaanxi, Gansu, and Inner Mongolia—have been lifted out of poverty. The farmers' incomes doubled, employment diversified, and the degraded environment was revitalized.

The project encouraged the natural regeneration of grasslands, trees, and shrub cover on previously cultivated slopes. Replanting and bans on free-range grazing increased perennial vegetation cover from 17 to 34 percent. This substantial increase in vegetation sequestered significant amounts of CO_2 into the vegetation and soil. Sedimentation of waterways was dramatically reduced by more than 100 million tons each year, reducing flooding risks. Prior to the project's implementation, frequent droughts caused crops cultivated on slopes to fail, requiring the government to provide emergency food aid. Terracing increased average yields and reduced yield variability by stopping erosion and capturing rainfall.

As many as 20 million people are benefitting from the impacts of this project's principles as they've been scaled up to regenerate other parts of China.[8] Inspired by the incredible success of the project, John realized there was a global need to regenerate degraded ecosystems. He founded the Ecosystem Restoration Communities (ERC) movement in 2016.

"Engaging large numbers of people immediately and directly to restore and revegetate degraded lands all over the world is the best way we have to sequester carbon and makes it possible for everyone to participate in mitigation and adaptation to climate change," says John.

In 2017, John hosted Ronnie, Ercilia Sahores, and me at Camp Altiplano, the first ERC project located on the high steppe of the Murcia region in southeastern Spain. The self-sufficient and off-grid community is based in an old village that had been largely abandoned due to the previous degenerative agricultural practices that impoverished the village farmers resulting in their flight to the cities. This is a classic story everywhere—the exodus of farmers due to poverty.

The camp is also supporting the local Alvelal collective to regenerate 620,000 hectares (1.5 million acres) of the degraded landscape. Camp Altiplano envisions a healthy, resilient regional ecosystem comprising a mosaic of productive and biodiverse farms, natural areas, and a regenerated economy. More than 2,000 people worldwide have participated through courses and volunteering programs. Since the initiative's inception, 20,000 native trees and bushes have been planted, with a 70 percent survival rate. In 2022, Altiplano harvested its first almonds after planting these trees in 2017, a momentous occasion for their regenerative almond agroforestry system. (I was so impressed with the potential of the Ecosystem Restoration Communities that I agreed to join the advisory committee and continue to be actively involved.)

By 2023, John's vision had grown to 56 communities covering 23,500 acres (9,400 hectares) on all six arable continents. Approximately 24,367 people have joined these communities, participating in regenerating our planet by planting over 3,250 trees and plants. These Ecosystem Restoration Communities strengthen localities and help restore their land and the biological life that stems from it. They are a straightforward way to fight poverty, famine, climate change, loss of freshwater resources, desertification, and biodiversity loss. Regenerative agricultural techniques such as permaculture, agroecology, and organic agriculture are implemented in the communities. They bring degraded areas back to life, enabling communities to benefit from a regenerated landscape.

CHAPTER FIVE

Business as Usual Is Not an Option

According to the United Nations Convention to Combat Desertification (UNCCD), industrial agriculture has altered the face of the planet more than any other human activity. It is responsible for 80 percent of deforestation and 70 percent of freshwater use, and is the most significant cause of terrestrial biodiversity loss. The industrial-agricultural revolution of the last century has neglected soil health and belowground biodiversity, especially the soil microbiome (the ultimate source of most of our food) due to its focus on toxic synthetic chemicals as the basis of increasing production.

The destruction of forests and diverse ecosystems primarily for commodity production generates the bulk of CO_2 emissions associated with land use change, nitrous oxides from synthetic chemical fertilizers, the methane emitted by factory farms, and intense animal production. These are all significant sources of greenhouse gas emissions.[1]

Soil carbon is the largest pool of carbon after the oceans. The soil holds almost three times as much carbon as the atmosphere, forests, and ecosystems combined.[2] Degenerative land use is oxidizing this organic carbon into CO_2. The loss of soil carbon through degenerative farming practices has been underestimated in its contribution to atmospheric greenhouse gases. Oxidation of soil carbon is caused by excessive tillage, bare soil, and erosion. Synthetic nitrogen fertilizers stimulate the types of microbes that consume soil carbon and turn it into CO_2. Research shows that it is a considerable contributor to the CO_2 in the atmosphere.

A recent study published by Kenneth Skrable, George Chabot, and Clayton French analyzed the change in the proportions of carbon-14 (C-14) in the atmosphere and raised doubts that the increase in CO_2 is mainly the result of burning of fossil fuels. "Our results show that the percentage of the total CO_2 due to the use of fossil fuels from 1750 to 2018 increased from 0 percent in 1750 to 12 percent in 2018, much too low to be the cause of global warming," the authors state.[3] This study, naturally, is very contentious; however, it is an essential part of the debate on how we manage the causes of climate change. The current focus is mainly on reducing fossil fuel use, methane production from ruminants, and scaling up renewable energy. The research shows this approach is highly problematic.

Instead, the authors propose that a large percentage of the increase in CO_2 in the atmosphere since 1750, from 280 ppm to over 400 ppm, comes from living carbon sources, not fossil fuels. These sources are obviously from clearing forests and the loss of soil carbon. About 1.5 billion hectares (3.7 billion acres) of forest have been cleared since 1750, the beginning of the Industrial Revolution. That's an area 1.5 times the size of the United States. This loss of forests has made, and continues to make, a massive contribution to the current CO_2 levels. These forests played an essential role in drawing down CO_2 through photosynthesis. Not only has this drawdown capacity been lost, but the degradation of all the organic biomass, 95 percent of which is cellulose, was oxidized into CO_2 and released into the atmosphere and oceans.

As I described in chapter 2, forests and rangelands are key in influencing local, regional, and transcontinental climates by creating much of their rain through transpiration. Transpiration of water vapor from plants is essential for correcting water vapor pressure deficits (VPDs) to stop the land from drying out and to ensure there is enough water vapor to induce rain. Transpiration from plants cools the atmosphere, reducing temperatures. Plants also shade the ground from direct sunlight, cooling the soil, and further reducing VPDs.

The Importance of Soil Organic Matter for Water Retention

Soil organic matter is key to the ability of soils to capture and retain rainfall. There is a strong relationship between the levels of soil organic matter and the amount of water that can be stored in the root zone of the soil. Different soil types will hold different volumes of water when they have the same organic matter levels due to pore spaces, specific soil density, and a range of other variables. Sandy soils, as a rule, hold less water than clay soils. Below, you'll find estimates for the volume of water that can be captured from rain and stored at the root zone in relation to the percentage of soil organic matter in the first six inches of soil per acre.[4] One percent is common in much of Africa, Asia, and parts of Latin America. Five percent was common in presettlement soils globally.

1 percent SOM = 16,640 gallons per acre
(160,000 liters per hectare)
2 percent SOM = 33,280 gallons per acre
(320,000 liters per hectare)
3 percent SOM = 49,920 gallons per acre
(480,000 liters per hectare)

4 percent SOM = 66,560 gallons per acre
(640,000 liters per hectare)
5 percent SOM = 83,200 gallons per acre
(800,000 liters per hectare)
6 percent SOM = 99,840 gallons per acre
(960,000 liters per hectare)

There is a large difference in the amount of rainfall that can be captured and stored between the current soil organic matter (SOM) level in most traditional farms in Asia, Africa, and Australia and a good regenerative organic farm with reasonable levels of SOM. This is one of the reasons why organic farms do better in times of low rainfall and drought.

Soil organic matter increases the soil's capacity to retain moisture, which aids in plant transpiration, evaporative cooling, air moisture, and local rainfall. In a 2023 paper, researchers showed that during wet and warm conditions, heat from the soil released into the air cools soil temperatures. This heat increases atmospheric water content, leading to more rainfall, which is usually associated with increased soil water content. "Our findings further support the importance of the soil moisture–temperature feedback for the evolution of hot spells in a warming climate," the authors wrote. The research also confirmed previous research showing that a lack of adequate soil moisture creates VPD. This important finding shows how VPD creates local heat waves.

"The fast increase in hot soil extremes is highly related to soil water loss, but other factors such as land cover change and land management could be affecting the different trends between air and soil hot extremes via changes in soil water content," the researchers wrote.[5]

This explains why clearing ecosystems in the Amazon, Borneo, Sumatra, Central America, and Central Africa has led to more frequent and severe droughts: the loss of SOM and forest cover heats up the soil and reduces soil moisture, causing hot, dry conditions. Hot air holds more moisture, so the rainfall is more torrential, often causing damaging floods when it rains. The key to moderating the climate is high levels of SOM to capture and retain rainfall to cool the climate.

In chapter 2, you may recall that I discussed how plants use the energy of sunlight to combine carbon dioxide and water to produce glucose in photosynthesis. Glucose is a primary energy source for the cells of most living organisms. It is a molecule of life. Ten to 40 percent is secreted into the soil through the roots while plants grow, feeding the soil microbiome.[6] The soil microbiome builds SOM.

In other words, the more plants and trees photosynthesize, the more glucose there is to feed the soil microbiome and, consequently, the bacteria's and fungi's ability to increase SOM. The key to maximizing SOM is to maximize photosynthesizing plants. Dead plants and bare soil do not photosynthesize.

Historical Perspectives

Up to the mid-nineteenth century, soil organic matter was referred to as "black vegetable mould" and was considered vital to soil fertility. In the 1840s, this concept lost favor when the two leading agronomists of the time, Count Justus von Liebig in Germany and Sir Bennet Lawes in the UK, conducted experiments showing that three macronutrients—nitrate, phosphorus, and potassium (NPK)—had the most impact on the growth and yields of crops. For the next 180 years, the emphasis remained squarely on soluble synthetic NPK fertilizers, with the soil being treated mainly as a medium that stopped the plants from falling over. SOM and the concept of soil health were de-emphasized from agricultural practices and only began to slowly reemerge in the twentieth century, thanks to the persistence of those in the organic agriculture movement.

It is impossible to get accurate data on how much atmospheric CO_2 is the result of lost SOM since very little data was recorded before primary ecosystems such as rangelands and forests were degraded by degenerative agricultural practices.

Australia was the last continent to be colonized by Europeans, the land forcibly taken from the Indigenous Australians, its traditional owners, and converted to European agricultural systems. From 1839 to 1843, Sir Paweł Edmund Strzelecki analyzed the soils of early agricultural farms in the present-day Australian states of New South Wales, Victoria, and Tasmania and found that the SOM ranged from 50 percent to a low of 5 percent, with an average of 33 percent.[7]

Strzelecki was a chemist who analyzed soils in Europe, North America, Central America, South America, the Pacific Islands, New Zealand, and Indonesia before coming to Australia. He stated, "In their external character, the soils of New South Wales and Van Diemen's Land [Tasmania] are, nevertheless, alike; particularly those which are as yet untouched by the hand of man, and which possess, in both colonies, the same degree of softness, coherence, and porosity common to all virgin soils, together with a low specific gravity, and a proportion of organic to inorganic matter amounting to a third, and in one instance to a half, of the whole quantity."[8]

Strzelecki documented the rapid decline of Australian soils once they were subjected to European agricultural systems. Numerous other early colonists, such as Sir Thomas Mitchell and Edward Curr, documented this rapid decline in Australia, as well; today, most agricultural soils in Australia have less than 1 percent SOM and are deficient in many key minerals. As Strzelecki presciently observed, "From the circumstance that, on the first introduction of tillage and grazing, the analysis of soils in particular fields was not performed, and the chemical nature of the first seed and the first crop not ascertained, it is difficult to determine the precise extent of this change, as regards soils. Judging, however, from the constituents of those untouched by the hand of man, the soils under tillage or pasturage have deteriorated in an agricultural point of view, having lost in salts and alkalies."

In the 1840s, Strzelecki clearly understood how degenerating soils caused local climate change, observing the rapid deterioration of soils when subject to tillage and grazing: "Furthermore, they have deteriorated in a climatic point of view, as their power of absorbing moisture from the atmosphere has been curtailed, and that of absorbing solar heat has increased; while that of retaining heat, during terrestrial radiation, has decreased." Strzelecki's observation is a clear example of what we now call vapor pressure deficit (VPD).

The science and observations used by Strzelecki in the 1830s and '40s have been overlooked for 180 years and only recently have been independently rediscovered in scientific research. A paper recently published in the journal *Nature Climate Change* showed how soil temperatures influence air temperature, "increasing soil evaporation, which may further dry and warm the soil highlighting the contribution of soil moisture–temperature feedback to the evolution of hot extremes in a warming climate."[9] The researchers demonstrated how bare soil absorbed large amounts of heat that were radiated back into the atmosphere, writing: "When the soil surface is warmer than the air above the surface, there is a heat exchange from the soil to the lower atmosphere. . . . This release of heat can contribute to the intensification and spreading of air hot extremes and heatwaves." This study underscores the importance of ground cover such as tree canopies and grasslands in shading the soil to prevent direct sunlight from heating it and SOM to capture rainfall to increase moisture levels.

Strzelecki documented that the loss of SOM through overgrazing and tillage led to the aridification of New South Wales:

> *The pasture, however, here, as in the foreground of the colony, began to diminish: the occasional burnings which were from time to time resorted to in order either to ameliorate the pasture, or to produce a new growth from the roots of the grasses, did but accelerate the slowly but evidently approaching evils. Dews began to be scarce, and rain still more so: one year of drought was followed by another; and, in the summer of 1838, the whole country of New South Wales between Sydney and Wellington, the Upper and Lower Hunter River, Liverpool Plains, Argyleshire, [etc.], presented, with very few exceptions, a naked surface, without any perceptible pasture upon it, for numerous half-starved flocks.*

This is still the case for large parts of Australia during the drier periods. The rich, diverse grasslands on soft, rich, moisture-holding soils were largely destroyed within a few years of colonization, creating VPDs that dried out and heated the landscape. The eroded and hardened ground no longer efficiently absorbed the rain, which ran off further, eroding it or evaporating it. Australian agriculture is now a continuous cycle of heavy rains causing damage followed by droughts due to the loss of soil organic matter, which has reduced the ability of the soil to capture rain and store it for the drier seasons.

North America, like Australia, had large regions that had never been subject to tillage cropping systems before colonization. Depending on the climate, soil type, and rainfall, virgin soils in the United States ranged from a low of 3 percent to a high of 12 percent soil organic matter. Many experts estimate that this has declined by 50 percent in most regions due to damaging tillage and poor grazing practices. Over 50 percent of the deep prairies' topsoils are estimated to have been lost to erosion. They have been completely lost in many places, with only the subsoils remaining.

The Dust Bowl is the most dramatic example of this, greatly damaging the American and Canadian prairies' ecology, agriculture, and communities during the 1930s. While historians tend to blame natural and man-made factors such as severe drought and unsuitable plowing as the main cause, the main cause was actually the soil's inability to capture and store rain and hold the soil together due to degenerative farming practices. This resulted in VPD that heated up the landscape, lowered rainfall, and dried out the soils. The soil, previously many yards deep with rich organic matter, turned into dust and blew away.

Indigenous Management Practices

The deep, rich soils of the prairies, savannas, and steppes were created through thousands of years of traditional ecological management systems by the Indigenous owners of these lands. Plains Indians used fire to develop fresh pasture. The buffalo followed them in order to eat the soft new green shoots. The Aboriginal Australians did the same, using fire to burn areas of grasslands in seasonal cycles so that kangaroos and emus would have the fresh green pick. The animals would naturally move to these areas. This is how people moved and managed their livestock without fences. These management systems built the deep, rich, fertile soils with the prairies' and savannas' highly diverse plant and animal species. The colonists from Europe quickly destroyed these diverse systems. Most had and still have no understanding of or respect for these ecologically complex, human-managed agroecosystems.

As stated previously, most of the deserts in North Africa, the Middle East, and Asia were created by continuous grazing that never allowed the plants or pastures to rest and regenerate. The trees, shrubs, herbs, flowers, and grasses were eaten out to the bare, eroded ground. Previously, traditional pastoral communities such as the Masai in East Africa, the Bedouin in the Middle East, the Kazakhs and Mongols in Asia, the Plains Indians in the United States and Canada, and Aboriginal Australians rotated their herds of native animals and did not return them to previously grazed pastures until those pastures had fully recovered. These systems have been continuously disrupted by "civilizations" based on intense sedentary cropping and grazing systems that unsustainably degraded the environment and created deserts, leading to collapse.

It was similar in Africa and South America. The savannas of Africa, such as the Serengeti, supported billions of grazing animals for millions of years. Traditional owners, such as the Masai, have grazed their herds of diverse species there for tens of thousands of years in harmony with the incredible diversity of plants and animals. This was disrupted by European colonialism. It is now threatened by land grabs and disruptions in the annual migrations of these species.

In the Pampas, the grasslands of South America, the situation is worse, with the land being transformed into vast monocultures of GMO soy and corn, sprayed with the highly toxic glyphosate that kills everything except for the crop. Further north, the rich, biodiverse tropical forests of the Mato Grosso and the Amazon are being cleared for monoculture crops. At the time of writing, the Amazon is enduring the worst drought in recorded history, with water levels dropping to record lows. Brazil has broken all winter and spring temperature records. Forests generate most of the rainfall in the Amazon and Mato Grosso. The clearing and burning for agriculture have created massive areas of VPD, prompting droughts. When the rains come, they are often torrential, causing destructive floods due to heat increasing the water-holding capacity of the atmosphere.

The worst thing about this destruction of primary ecosystems is that it's not even done in the name of nourishing the food-insecure people across the globe. Industrial commodity crops are shipped in large cargo boats to western Europe and China to feed animals in cruel concentrated animal feeding operations (CAFOs). For every 10 tons of protein fed to these cruelly confined animals, the result is 1 ton of animal protein. These commodities feed the world's wealthiest people with the highest levels of obesity and chronic diseases, not those who are starving or suffering from malnutrition. This industrial system is the most perverse, inefficient, environmentally destructive, and cruel way to produce food. It must be banned for many reasons, including its massive contribution to climate change.

Many of the most fertile and productive regions in Asia, Africa, and the Middle East had been turned into deserts and lost all their soil organic matter thousands of years ago. Other regions, such as the Mayan regions of the Yucatán Peninsula and the Khmer civilization in Cambodia, Thailand, Laos, and Vietnam, collapsed because they destroyed the soils hundreds of years ago. Allowing their forests to regenerate also regenerated their soils and climates.

Without records of the original SOM levels, it is impossible to calculate the true loss. But Australian agricultural soils had an average of 33 percent SOM (with the lowest level of

5 percent) prior to colonization based on the only chemist who measured it. Records show that 12 percent SOM was common in the United States before colonization reduced it by 50 percent. Now, most agricultural soils have between 0.5 to 3 percent SOM, with most soils in Africa, Asia, Australia, and Mediterranean Europe having less than 1 percent. My experience in finding the soils from remnant forest and grassland ecosystems is that previously most of these farming areas started with around 6 percent or higher SOM.

According to Dr. Christine Jones, "An increase of 1 percent in the level of soil carbon in the 0–30 cm soil profile equates to sequestration of 154 tons CO_2/ha with an average bulk density of 1.4 g/cm^3."[10] Australia has 1,065 million acres (426 million hectares) of agricultural land. Assuming a low level of 6 percent SOM prior to colonization and a current average of less than 1 percent SOM translates to a loss of 5 percent SOM. This means a conservative estimate of 328 gigatons of CO_2 emitted into the atmosphere. The United States has 900 million acres (360 million hectares) of agricultural land and had an average of 12 percent SOM at colonization, with an estimated loss of half of it, or 6 percent. This translates to a conservative estimate of 332 gigatons of CO_2 emitted into the atmosphere.

Australia and the United States alone are responsible for 660 billion tons of atmospheric CO_2 from the loss of soil organic matter. This shows that on a global scale, trillions of tons of CO_2 have been lost from the soil and ended up in the atmosphere. This huge number is ignored in the current greenhouse gas estimates.

Clearing 1.5 billion hectares (3.7 billion acres) of forest since the beginning of the Industrial Revolution and converting them into agriculture has resulted in a massive decline in SOM and increases in atmospheric CO_2. Agriculture, forest, and biodiversity management must change. They are a significant cause of climate change, possibly greater than fossil fuels. It is time to regenerate our soils, agricultural and natural ecosystems, and climates by scaling up regenerative agriculture.

Multifunctional Benefits

Scaling up regenerative agriculture can reverse climate change, increase biodiversity, improve water capture and retention, stop soil loss, be more profitable for farmers and ranchers, and nourish the world with high yields of healthy, nontoxic food. (Most of the published studies that support these assertions are based on organic agriculture, as it is the oldest and most widespread of regenerative systems.)

A comprehensive body of published studies shows that building up soil organic matter through regenerative organic agriculture can deliver many benefits, including higher water absorption, erosion resistance, higher drought yields, and adequate soil nitrogen without the need for synthetic chemical fertilizers. Of great significance, as I argued in chapter 2, is that the worldwide adoption of these regenerative organic practices could sequester enough CO_2 to keep the predicted global temperature rise to less than 2°C (3.6°F), thereby ameliorating climate change.

A large body of scientific evidence based on hundreds of published peer-reviewed scientific studies shows the harm that pesticides used in industrial agriculture are causing to human and environmental health. Science shows that there are no safe levels of pesticides for fetuses, young children, and youths going through puberty. The science shows that the smallest amounts can cause significant lifetime health problems.[11]

The Rodale Institute's 40-Year Report on their Farming Systems Trial should end the myth of industrial agriculture and its toxic, GMO-herbicide, no-till systems. Rodale's scientific trials clearly demonstrate that these degenerative systems are inferior to regenerative organic agriculture on every key criterion.[12] Indeed, there are many compounding and multifunctional benefits to regenerative agriculture systems that extend far beyond climate, crop yields, and a nutritionally adequate food supply.

Higher Net Incomes

A viable income is an essential part of farm sustainability. Published studies comparing the income of organic and industrial farming systems found that the net incomes are similar, but gave an edge to best-practice organic systems.[13] A 2013 United Nations Food and Agriculture Organization (FAO) study analyzed more than 50 economic studies and found:

> *In the majority of cases, organic systems are more profitable than non-organic systems. Higher market prices and premiums, or lower production costs, or a combination of the two generally result in higher relative profits from organic agriculture in developed countries. The same conclusion can be drawn from studies in developing countries, but there, higher yields combined with high premiums are the underlying causes of their relatively greater profitability.*[14]

The report, published by the United Nations Environment Program (UNEP) and the United Nations Conference on Trade and Development (UNCTAD), found that not only did organic production increase the amount of food production, but it also gave farmers access to premium value markets and the ability to use the additional income to pay for education, health care, adequate housing, and achieve relative prosperity.[15] Similarly, Iowa State University's Long-Term Agroecosystem Research (LTAR) project found that, cost-wise, on average, the revenue of agroecological organic crops was twice that of conventional crops due to the savings from the non-utilization of chemical fertilizers and pesticides.[16]

Water Efficiency

Research shows that regenerative organic systems use water more efficiently due to better soil structure and higher levels of humus and other organic matter compounds.[17] During the Rodale Farm Systems Trial, Lotter and colleagues collected data over the course of 10 years and found that regenerative organic systems improved the soils' water-holding capacity, infiltration rate, and water capture efficiency. The regenerative organic soils averaged 13 percent higher water content than industrial systems.[18]

The more porous structure of organically treated soil allows rainwater to quickly penetrate the soil, resulting in less water loss from runoff and higher levels of water capture. This was particularly evident during the two days of torrential downpours from Hurricane Floyd in September 1999, when organic systems captured around double the water of industrial systems.[19] This ability to capture more water is due to the strong relationship between the levels of soil organic matter and the amount of water that can be stored in the root zone of the soil. As we saw earlier in this chapter, every 1 percent increase in SOM increases water storage and capture by 16,640 gallons per acre.

This is one of the reasons why regenerative organic farms do better in times of low rainfall and drought, which is a significant piece of information as most of the world's farming systems are rain-fed. The world does not have the resources to irrigate all of its agricultural lands. Nor should we start damming the world's watercourses, pumping from all the underground aquifers, and building millions of miles of channels; this would be an unprecedented environmental disaster. Improving the efficiency of rain-fed agricultural systems through regenerative practices is the most efficient, cost-effective, environmentally sustainable, and practical solution to ensure reliable food production during times of increasing weather extremes.

Preventing Soil Erosion and Loss

Long-term scientific trials (known as the DOK Trials) conducted by the Research Institute of Organic Agriculture (FiBL) in Switzerland compared organic, biodynamic, and conventional systems and found that organic systems were more resistant to erosion and better at capturing water. This is consistent with many other comparison studies showing that organic systems have less soil loss due to better soil structure and higher levels of organic matter.[20]

Long-term research conducted in Washington State by Professor John Reganold and colleagues compared the long-term effects (since 1948) of organic and conventional farming on selected properties of the same soil. They found that "the organically farmed soil had significantly higher organic matter content, thicker topsoil depth, higher polysaccharide content, lower modulus of rupture, and less soil erosion than the conventionally farmed soil. This study indicates that, in the long term, the organic farming system was more effective than the conventional farming system in reducing soil erosion and, therefore, in maintaining soil productivity."[21]

Humus, a key component of soil organic matter, is one of the main reasons for the ability of organic soils to be more stable and to hold more water—up to 30 times its own weight in water! Being a "sticky polymer," it binds the soil particles together, giving greater resistance to water and wind erosion. It also gives the soil structure and stability and holds many of the nutrients that plants need to grow well.

Drought Resistance

Published studies show that organic farming systems are more resilient to extreme weather conditions and can produce higher yields than conventional farming systems in such conditions.[22] For instance, the Wisconsin Integrated Cropping Systems Trials found that organic yields were higher in drought years and the same as conventional systems in normal weather years.[23]

The Rodale Institute's 40-Year Report on their Farming Systems Trial showed that organic maize yields have been 31 percent higher than conventional/industrial farming systems in drought years.[24] The researchers attribute higher yields in the dry years to the organic soil's ability to absorb rainfall better due to higher levels of SOM. The friability of such soils allows them to better capture and store rainwater, which can then be used for crops.

Scaling Up

There are numerous examples of regenerative agricultural systems that demonstrate we can successfully increase production to nourish the food-insecure, end poverty, and restore ecosystems, hydrology, and the climate. The key is maximizing photosynthesis to draw down CO_2 and increase soil organic matter, thereby maximizing the capture and retention of rainfall. This will increase transpiration, cool the atmosphere, and reduce VPD. Indeed, regenerating the planet by scaling up a diversity of these nature-based solutions rather than destructive and toxic industrial-scale constructions is our only viable future.

Ronnie and I spent so much time and effort thinking, discussing, and mapping out how this can be achieved. For Ronnie, the most significant project was scaling up the agave agroforestry system through the Billion Agave Project. Ronnie explains the multiple benefits along with the economics of doing this in chapter 6.

CHAPTER SIX

Agave Power

Agave, from the Greek word αγαυή, meaning "noble" or "admirable," is a common perennial desert succulent plant, with thick fleshy leaves and sharp thorns. Agave plants evolved originally in Mexico, and were utilized by Indigenous people for thousands of years as a source of food, fiber, medicine, soap, paper, firewood, and beverages (both alcoholic and nonalcoholic).[1] Agaves are also found today in the hot, arid, and semiarid drylands and tropical regions of Central America, the Caribbean, the Southwestern United States, South America, Africa, Oceania, and Asia.

Agaves spread naturally through the Americas, but were also transported by Europeans to their overseas colonies in Africa and Asia. Agave is best known for producing textiles (henequen and sisal) from its fibrous leaves; a syrup sweetener (jarabe); alcoholic beverages—tequila, *pulque*, and mescal—from its sizeable root stem, or *piña*; inulin, a prebiotic nutritional supplement from its *piña*; pet food supplements from the bagasse or leftover pulp after the liquid is removed; and bioethanol. It also serves as an ornamental plant.

Agave's several hundred different varieties are found growing on approximately 20 percent of the Earth's surface, often growing in the same desertified, degraded cropland or rangeland areas as nitrogen-fixing, deep-rooted trees or shrubs such as mesquite, acacia, or leucaena. Agaves can tolerate intense heat (up to 135°F or 57°C) and will readily grow in drylands or semi-desert landscapes where there is a minimum annual rainfall of approximately 10 inches or 250 mm. In colder areas agaves can survive temperatures of 14°F (–10°C).

The United Nations Convention to Combat Desertification (UNCCD) estimates that arid and semiarid lands make up 41.3 percent of the Earth's land surface, including 15 percent of Latin America (60 percent of Mexico), 66 percent of Africa, 40 percent of Asia, and 24 percent of Europe.

Agaves Need No Irrigation

Agaves basically require no irrigation, efficiently storing seasonal rainfall and moisture from the air in their thick thorny leaves (*pencas*) and stem or heart (*piña*) utilizing their Crassulacean acid metabolism (CAM) photosynthetic pathway, which enables the plant to grow and produce significant amounts of biomass, even under conditions of severely restricted water availability and prolonged droughts. Agaves reproduce by putting out shoots or *hijuelos* alongside the mother plant—approximately 3 to 4 per year—or through seeds, if the plant is allowed to flower at the end of its 8- to 13-year (or more) lifespan.

A number of agave varieties appropriate for drylands agroforestry (*salmiana*, *americana*, *mapisaga*) readily grow into large plants, reaching a weight of up to 650 kilograms (1,400 pounds) in the space of 8 to 13 years.

Agaves are among the world's top 15 plants or trees in terms of drawing down large amounts of carbon dioxide from the atmosphere, producing plant biomass, and exuding liquid carbon through its root structure in the soil.[2] Appropriate varieties of agave are capable of producing large amounts of fermented animal silage per hectare (up to 100 tons of silage per hectare, or 40 tons per acre) on a continuous basis from year three, increasing proportionately as the plant matures. In addition, the water utilization of agaves (and other desert-adapted CAM plants), which it can obtain solely from rainfall, is typically 4 to 12 times more efficient than other plants and trees, with average water demand approximately 6 times lower.

Agave-Based Agroforestry and Silage Production

Agave's nitrogen-fixing, deep-rooted companion trees or shrubs such as mesquite and acacia have adapted to survive in these same dryland environments as well. From an environmental, soil health, and carbon-sequestering perspective, agaves should be cultivated and interplanted, not as a monoculture—as is commonly done with *agave azul* (the blue agave species) on tequila plantations in Mexico (often 3,000 to 4,000 plants per hectare, or 1,215 to 1,600 plants per acre), or on mescal monocultures (with the *espadin* variety)—but as a polyculture.

In a polyculture agroforestry system, several varieties of agave are interspersed with native nitrogen-fixing trees or shrubs (such as mesquite or acacia), native vegetation, pasture grass, and cover crops, which fix the nitrogen and nutrients into the soil that the agave needs to draw upon in order to grow and produce significant amounts of biomass / animal forage. If grown as a polyculture, agaves and their companion trees and shrubs can be cultivated on a continuous basis, producing large amounts of biomass for silage and sequestering significant amounts of carbon aboveground and belowground on a continuous basis, without depleting soil fertility (especially nitrogen) or biodiversity. In our Vía Orgánica research farm in San Miguel de Allende, our practice is to leave as much prior biodiversity and ground cover as possible.

In addition to these polyculture practices, planned rotational grazing on these agroforestry pastures, once established, not only provides significant forage for livestock, but done properly (neither overgrazing nor under-grazing) further improves or regenerates the soil, eliminating dead grasses and invasive species, facilitating water infiltration (in part through ground disturbance from hoof prints), concentrating animal manure and urine, and increasing soil organic matter, soil carbon, biodiversity, and fertility.

An Ignored and Denigrated Species

Although agave is a plant that grows prolifically in some of the harshest climates in the world, in recent times this plant has been largely ignored, if not outright denigrated. Apart from producing alcoholic beverages, agaves are often considered a plant and livestock pest, along with its thorny, nitrogen-fixing, leguminous companion trees or shrubs such as mesquite and acacia.

But now, the Flores González brothers' new agave-based agroforestry and holistic livestock management system—which utilizes basic ecosystem restoration techniques, permaculture design, and silage production using anaerobic fermentation—is changing the image of agave and their companion trees. This agave-powered agroforestry and livestock management system is demonstrating that native desert plants, long overlooked, cultivated as part of an agroforestry system, have the potential to regenerate the drylands, provide large amounts of inexpensive animal feed, and increase essential forage for grazing animals, in the process taking pressure off overgrazed rangelands, improving animal health, rehydrating parched soils, and alleviating rural poverty.

Moving beyond conventional monoculture and chemical-intensive farm practices, and combining the traditional Indigenous knowledge of native desert plants and natural fermentation, this innovative group of Mexico-based farmers have learned how to reforest and green their drylands, all without the use of irrigation or expensive and toxic agricultural inputs.

Agroforestry Design

Organic farmers and researchers have created this new agroforestry program by densely planting, pruning, and intercropping high-biomass, high-forage producing species of agaves (an average of 2,000 per hectare, or 810 per acre) among preexisting deep-rooted, nitrogen-fixing tree or shrub species (400 per hectare) such as mesquite and acacia, or alongside transplanted tree saplings. For reforestation and biodiversity in the agave/mesquite agroforestry system, the Vía Orgánica research farm in San Miguel has developed *acodos* or air-layered clones of mesquite trees. These mesquite *acodos* are essentially mesquite branches from mature trees transformed into saplings planted into the ground, which, after one year of being watered, fertilized, and cared for in the greenhouse, can grow up to 2 meters (6 feet) tall. In comparison, a one-year-old mesquite sapling from seed is typically 35 centimeters (1 foot) tall.

Agaves (especially *salmiana*, *americana*, and *mapisaga*) naturally produce large amounts of plant leaves or *pencas* every year, which can then be pruned (20 percent per year starting in year three of its lower leaves), chopped up, fermented, and turned into silage. Agave's perennial silage production far exceeds most other forage production (most of which require irrigation and expensive chemical inputs) with three different varieties—*salmiana*, *americana*, and *mapisaga*—in various locations producing approximately 40 tons per acre (100 tons per hectare) of fermented silage, annually. The variety *crassispina*, valuable for its high-sugar *piña* content for mescal, produces slightly less than 50 percent of the *penca* biomass of the other three varieties (averaging 46.6 tons per year).

The Economics of Agave Silage

The fermented agave silage of the three most productive varieties has a considerable market value of $100 per ton (up to $4,000 per acre or $10,000 per hectare in gross income). This perennialized, polyculture system, in combination with rotational grazing, has the capacity to feed up to 40 sheep, lambs, or goats per acre per year (or 100 per hectare), producing a potential value-added net income (from meat, milk, cheese, and viscera) of $3,000 per acre ($7,500 per hectare). Once certified as organic, sheep and lamb production can substantially increase profitability per acre/hectare, especially if organic viscera (heart, liver, kidneys, and so on) are processed into freeze-dried nutritional supplements.

In addition, the agave heart or *piña*, with a market value of $150 per ton, harvested at the end of the agave plant's 8- to 13-year lifespan for mescal, *pulque*, inulin, or silage, can weigh 300 to 400 kilograms (660–880 pounds) in the three most productive varieties. Again, the *crassispina* variety has a much smaller *piña* (160 tons per 2,000 plants). The value of the *piña* from 2,000 agave plants for the *salmiana*, *americana*, and *mapisaga* varieties, harvested once, at the end of the plant's productive lifespan (approximately 10 years) has a market value of $52,500 per hectare (over 10 years, with 10 percent harvested annually), with the market value for inulin being considerably more.

Combining the market value of the *penca* and *piña* of the three most productive varieties we arrive at a total gross market value of $152,500 per hectare ($61,538 per acre) over ten years. Adding the value of the 72,000 *hijuelos* or shoots of 2,000 agave plants (each producing an average of 36 shoots or clones) with a value of 12 pesos or 60 cents USD per shoot, we get an additional $43,200 gross income over ten years. Total estimated gross income per hectare for *pencas* ($100,000), *piñas* ($52,500), and *hijuelos* ($43,200) over ten years will be $195,700, while expenses to establish and maintain the system are projected to be $16,047 per hectare over 10 years. As these numbers, even though approximate, indicate, this system has tremendous economic potential.

Pioneered by sheep and goat ranchers, Hacienda Zamarripa, in the municipality of San Luis de la Paz, Mexico, and then expanded and modified by organic farmers and researchers in San Miguel de Allende and other locations—the "Billion Agave Project," as the new movement calls itself—is starting to attract regional and even international attention from farmers, government officials, climate activists, and impact investors. One of the most exciting aspects of this new agroforestry system is its potential to be eventually established or replicated, not only across Mexico, but in a significant percentage of the world's arid and semiarid drylands, including major areas in Central America, the Caribbean, Latin America, the Southwestern United States, Asia, Oceania, and Africa.

Besides improving soils, regenerating ecosystems, and sequestering carbon, the economic impact of this agroforestry system appears to be a long-overdue game changer in terms of reducing and eliminating rural poverty. Currently 90 percent of Mexico's dryland farmers, 86 percent of whom do not have wells or irrigation, are unable to generate any profit whatsoever from farm production, according to government statistics. The average rural household income in Mexico is approximately $5,000–6,000 USD per year, derived overwhelmingly from off-farm employment and remittances or money sent home from Mexican immigrants working in the United States or Canada. Almost 50 percent of Mexicans, according to government statistics, are living in poverty or extreme poverty.

Deploying the Agave-Based Agroforestry System

The first step in deploying this agave-powered agroforestry and holistic livestock management system involves carrying out basic ecosystem restoration practices. Restoration is necessary given that most of the world's dryland areas suffer from degraded soils, erosion, low fertility, low soil organic carbon, and low rainfall retention.

Initial ecosystem restoration typically requires putting up fencing or repairing fencing for livestock control, constructing rock barriers (check dams) for erosion control, building up contoured rows and terracing, subsoiling (to break up hardpan soils), transplanting agaves of different varieties and ages (1,600–2,500 per hectare or 650–1,000 per acre), sowing pasture grasses, and (if not previously forested) transplanting mesquite or other nitrogen-fixing trees and shrubs (400 per hectare or 160 per acre). Depending on the management plan, not all agaves will be planted in the same year, but ideally the system will contain an equal division of 200 plants per year of each age (planted at ages 1–10) so as to stagger harvest times for the agave *piñas*, which are harvested, along with all the remaining leaves or *pencas*, at the end of the particular species' lifespan.

Planting is followed by no-till soil management, after initial subsoiling, and sowing pasture grasses and cover crops of legumes, meanwhile temporarily "resting" pasture (in other words, keeping animals out of overgrazed pastures or rangelands) long enough to allow regeneration of forage and survival of young agaves and tree seedlings. Following these initial steps of ecosystem restoration, planting agaves, and establishing sufficient tree cover, which can take up to five years, a complementary step is carefully implementing planned rotational grazing of sheep and goats or other livestock across these pasturelands and rangelands, at least during the four-month rainy season. This will utilize moveable solar fencing and/or shepherds and shepherd dogs (neither overgrazing nor under-grazing), and supplementing or even replacing pasture forage, especially during the eight- to nine-month dry season, with fermented agave silage (wet) and agave flour (dried) and other protein sources, such as alfalfa, legumes, mesquite flour, and moringa.

During the dry season many families will choose to keep the breeding stock and their young offspring on their smaller family parcels or paddocks, feeding them fermented silage (either agave or agave/mesquite pod mix) to keep them healthy throughout the dry season, when pasture grasses are severely limited. In the Hacienda Zamarripa model, lambs can thrive during their five- to seven-month lifespan (before the animal is sold for slaughter) on nothing but fermented forage and their mother's milk. On the Vía Orgánica research farm other livestock and poultry also consume fermented agave (often mixed with agave flour, mesquite flour, or other protein sources). At Vía Orgánica up to 80 percent of goats' and sheep's diets are fermented agave (up to 50 percent for pigs, and up to 20 percent for chickens, ducks, turkeys, geese, and rabbits).

By implementing these restoration and agave agroforestry practices, farmers and ranchers can begin to regenerate and rehydrate dryland landscapes, improve the health and productivity of their livestock, provide affordable food for their families, improve their livelihoods, and at the same time deliver valuable ecosystem services, reducing soil erosion, recharging water tables, and sequestering and storing large amounts of atmospheric carbon in plant biomass and soils, both aboveground and belowground.

A Revolutionary Innovation

The Flores González brothers' revolutionary innovation has been to turn a heretofore indigestible, but massive and accessible, source of fiber, biomass, water, and protein—the agave leaves or *pencas*—into a valuable animal feed, utilizing the natural process of anaerobic fermentation to transform the plant leaves' relatively indigestible saponin compounds into digestible carbohydrates and sugar. To raise protein levels in the silage, the Vía Orgánica research farm combines the wet agave silage (with 3 to 5 percent protein) with dry agave flour (9 percent protein) and/or other protein supplements such as legumes to achieve the desired protein levels.

To shred the *pencas* or *piñas* for wet silage, the Zamarripa pioneers have designed and built a relatively simple machine, either self-powered, hooked up to a tractor, or electric-driven, that can chop up the very tough pruned leaves of the agave, producing over 1 ton of wet silage per hour. After shredding the agave's leaves or *pencas* (into what looks like green coleslaw) they anaerobically ferment this wet silage in a closed container, such as a 5- or 50-gallon plastic container with a lid, removing as much oxygen as possible (by tapping it down) before closing the lid. A separate machine, fairly simple and inexpensive, can grind mesquite pods, beans, moringa plants, or dried agave leaves into a higher protein agave flour.

The fermented end-product, golden-colored after 30 days, and good for 30 months, is a nutritious but inexpensive silage or animal fodder, which costs approximately 1.5 Mexican pesos (or 7.5 cents USD) per kilogram (or 2.2 pounds) for fermented agave alone, or 3 pesos—for agave and mesquite/bean pods or agave flour together—per kilogram to produce. In San Miguel de Allende, the containers used during the initial experimental stage of the project cost $3 per unit for a 20-liter or 5-gallon plastic container or *cubeta* with a lid, with a lifespan of 25 uses or more before they must be recycled. Two hundred–liter reusable containers cost $60 per unit (new; $30 used) but will last considerably longer than the 20-liter containers. At Vía Orgánica, before feeding our animals (sheep, goats, pigs, chickens, ducks, turkeys, rabbits, burros) the silage, we add dried agave flour or other protein sources like beans or alfalfa to augment protein.

As the preliminary numbers indicate, fermenting agave silage, harvesting and processing the *pencas* alone, will provide significant value and profits per hectare for landowners and rural communities, such as Mexican *ejidos*, who deploy the agave agroforestry system at scale. In addition, Billion Agave Project researchers are now developing silage storage alternatives that will eliminate the necessity for the relatively expensive 20-liter or 200-liter plastic *cubetas*.

Regenerative Economics

The agave silage production system can provide the cash-strapped rancher or farmer with an alternative to having to purchase alfalfa (expensive at 20 to 40 cents USD per kilogram and water-intensive), hay (likewise expensive), or corn stalks (labor intensive and nutritionally deficient), especially during the dry season.

According to Dr. Juan Frias, lead scientist for Vía Orgánica, lambs or adult sheep readily convert 10 kilos of fermented agave silage (wet) into 1 kilo of body weight, half of which will be marketable as meat or viscera. At 1.5 to 3 pesos per kilo (7.5 to 15 cents per pound), this highly nutritious silage can eventually make the difference between poverty and a decent income for literally millions of the world's dryland small farmers and herders.

Typically, an adult sheep will consume 2 to 2.5 kilograms of silage every day, while a lamb of up to 5 months of age will consume 500 to 800 grams per day. Cattle will consume 10 times as much silage per day as sheep, approximately 20 to 25 kilograms per day. Under the agave system it costs approximately 20 pesos ($1) per pound in live weight to produce sheep and goats. These can be sold at market rates for nonorganic mutton or goats at 40 pesos ($2) per pound live weight. Certified organic lamb, mutton, or goat will bring in 25 to 50 percent more. In ongoing experiments in San Miguel de Allende, pigs and chickens have remained healthy and productive with fermented agave forage providing 15 to 50 percent of their diet, reducing feed costs considerably.

The bountiful harvest of this regenerative, high-biomass, high-carbon-sequestering system includes not only extremely low-cost, nutritious animal forage (up to 60 to 100 tons or more, depending on the variety per hectare per year of fermented silage—starting in years 3 to 5, averaged out over 10 years), but also high-quality organic lamb, mutton, cheese, milk, *aquamiel* (agave sap), *pulque*, inulin, and mescal, all produced organically with no synthetic chemicals or pesticides whatsoever, at affordable prices, with excess agave biomass fiber, and bagasse available for textiles, pet food, compost, biochar, construction materials, and bioethanol.

The Bottom Line

It is necessary to have a strong economic incentive in order to motivate a critical mass of impoverished farmers and ranchers struggling to make a living in the degraded drylands of Mexico, or in any of the arid and semiarid areas in the world, to adopt this system. There absolutely must be economic rewards, both short-term and long-term, in terms of farm income, if we expect rapid adoption of this system. Fortunately, the agave/mesquite agroforestry system provides this, starting in year three and steadily increasing each year thereafter, producing significant amounts of low-cost silage to feed farmers' livestock and a steady and growing revenue stream from selling their surplus *pencas*, *piñas*, and silage from their farm or communal lands (*ejidos*).

Given that most of these dryland farmers have little or no operating capital, there needs to be a system to provide financing—such as loans, grants, and ecosystem credits—and technical assistance to deploy this regenerative system and maintain it over the crucial 5- to 10-year initiation period. Based upon a decade of implementation and experimentation, we estimate that this agave agroforestry system will cost approximately $1,600 per year per hectare to establish and maintain, averaged out over a 10-year period. By the fifth year, however, this system should be able to pay out initial operating loans (upfront costs in the first five years are much higher than in successive years) and begin to generate a net profit.

The overwhelming majority of Mexican dryland farmers have no wells for irrigation (86 percent) and make little or no money (90 percent) from their subsistence agriculture practices of raising corn, beans, squash, and livestock. Although the majority of rural smallholders are low-income or impoverished, they do typically own their own (family or self-built) houses and farm sheds or buildings as well as hold title to their own parcels of land, typically 5 hectares (12 acres) or less, as well as their livestock. And beyond their individual parcels, 3 million Mexican families are also joint owners of communal lands or *ejidos*, which constitute 56 percent of total national agricultural lands (103 million hectares or 254 million acres).

Rising Out of Poverty

Unfortunately, most of the lands belonging to Mexico's 28,000 communal landholding *ejidos* are arid or semiarid with no wells or irrigation. But being an *ejido* member does give a family access and communal grazing (plus some cultivation) rights to the typically overgrazed *ejido*. Some *ejidos*, including those in the drylands, are quite large, encompassing 12,000 hectares (30,000 acres) or more. In contrast to farmers in the United States or the rest of the world, most of these Mexican dryland farmers have little or no debt. For many, their bank account is their livestock, which they sell as necessary to pay for the ordinary household and personal expenses.

As noted earlier, most Mexican farmers today subsist on the income from off-farm jobs by family members, and remittances sent home from family members working in the United States or Canada. They understand firsthand that climate change and degraded soils are making it nearly impossible for them to grow their traditional *milpas* (corn, beans, and squash) during the rainy season or raise healthy livestock for family consumption and sales. Most are aware that their livestock often cost them as much (or more) labor and money to raise than their value for family subsistence or their value in the marketplace.

Mexico has a total of 2,400 *municipios* or counties located in 32 states. Small farmers are already cultivating wild or semi-wild agave plants to harvest the *piñas* to make *pulque* or mescal in 1,000 of these counties. These counties include many of the poorest communities in Mexico.

Most agave producers are not yet harvesting the agave leaves and *piñas* (up to 60 to 100 tons per year per hectare) in order to make large quantities of nutritious and inexpensive fermented livestock silage, however. Farmers in only a few areas—Hacienda Zamarripa in San Luis de la Paz, Vía Orgánica (and surrounding *ejidos*) in San Miguel de Allende, Guanajuato, the Mixteca region of Oaxaca, and *pulque* producers in Tlaxcala—are currently harvesting *pencas* to produce fermented silage for livestock. But as word spreads about the incredible value of *pencas* and the agave/mesquite agroforestry system developing across Mexico, farmers in most of the nation's *ejidos* and *municipios* will likely be interested in deploying this system in their areas, partly because there is currently no alternative.

With start-up financing, operating capital, and technical assistance, much of which can be farmer-to-farmer training, a critical mass of Mexican smallholders should be able to benefit enormously from establishing this agave-based agroforestry and livestock management system on their private parcels, and benefit even more by collectively deploying this system with their other *ejido* members on communal lands. With the ability to generate a net income up to $6,000 to $12,000 per year/per hectare of fermented agave silage (and lamb/sheep/livestock production) on their lands, low maintenance costs after initial deployment, and with production steadily increasing 3 to 5 years after implementation, this agave system has the potential to spread all across Mexico, as well as to all the arid and semiarid drylands of the world.

As tens of thousands, and eventually hundreds of thousands, of small farmers and farm families start to become self-sufficient in providing 100 percent of the feed and nutrition for their livestock, dryland farmers will have the opportunity to move out of poverty and regenerate household and rural community economies, restoring land fertility and essential ecosystem services at the same time.

The extraordinary characteristic of this agave agroforestry system is that it generates almost immediate rewards. Starting from seedlings or agave shoots (*hijuelos*) in year 3 of the 8- to 13-year lifespan, farmers can begin to prune and harvest approximately 20 percent of leaf biomass of the lower plant leaves or *pencas* every year and start to produce tons of nutritious fermented animal feed/silage. Individual agave leaves or *pencas* from a mature plant can weigh more than 20 kilos (45 pounds) each.

Because the system requires no synthetic inputs or chemicals, the meat, milk, or forage produced can readily be certified organic, likely increasing its wholesale value in the marketplace. In addition, the *piñas* from 2,000 agave plants (one hectare), with an average *piña* per plant of 300 to 400 kilograms, at a price of 3 pesos or 15 cents per kilogram, can generate a one-time revenue of $52,500 when all remaining leaves and stems are harvested.

Even though the agaves are completely harvested at the end of their lives, other agave seedlings or *hijuelos* (shoots) of various ages, which will have been steadily planted alongside them, will maintain biomass and silage production. In a hectare of 2,000 agave plants, approximately 72,000 *hijuelos* or new baby plants (averaging 36 per mother plant) will be produced over a 10-year period. These 72,000 baby plants (ready for transplanting) have a current market value over a 10-year period of 12 pesos (60 cents) each, or $43,200.

Financing the Agave-Based Agroforestry System

Although Mexico's dryland smallholders are typically debt-free, they are cash-poor. To establish and maintain this system, they need approximately $1,600 a year per hectare ($648 per acre) for a total cost over 10 years at $16,000 per hectare. Starting in year 5, each hectare should be generating $10,000 worth of fermented silage or *foraje* per year.

By year 5, farmers deploying the system will be generating enough income from silage production and livestock sales to pay off the entire 10-year loan. From this point on they will become economically self-sufficient, and in fact will have the opportunity to become moderately prosperous. Pressure to overgraze communal lands will decrease, as will the pressure on rural people to migrate to cities or to the US and Canada.

Meanwhile massive amounts of atmospheric carbon will have begun to be sequestered aboveground and belowground, enabling many of Mexico's 2,400 *municipalidades* to eventually reach net zero, and *net negative* carbon / greenhouse gas emissions. In addition, other ecosystem services will improve, including reduced topsoil erosion, more rainfall/water retention in soils, more soil organic matter, increased tree and shrub cover, increased biodiversity (aboveground and belowground), increased pollinator, bird, and wildlife habitat, restoration of grazing areas, and increased soil fertility.

Greening the Desert

In Mexico, where 60 percent of all farmlands or rangelands are arid or semiarid, this system has the capacity to sequester 100 percent of the nation's current greenhouse gas annual emissions (650 million tons of CO_2-eq per year) if deployed on approximately 23 percent of the nation's total lands (197 million hectares) over the next 10 years (so, on 45 million hectares or 111 million acres, with 2,000 agaves and 400 mesquites per hectare).

Communally-owned *ejido* lands alone account for more than 100 million hectares, while arid and semiarid lands (basically with no irrigation) amount to 120 million hectares. These may sound like large numbers, and indeed they are, but given the magnitude of the climate, environmental, and economic crisis, and the lack of any other practical solution, scaling up agave power and the agave agroforestry system offers perhaps the only short-term solution. The largest ecosystem restoration project in recent times has been the decade-long restoration of the Loess Plateau (1.5 million hectares) in north-central China in the 1990s. The current Great Green Wall project, stretching west to east across 30 nations in Africa, just below the Sahel desert, aims to regenerate 100 million hectares.

San Miguel de Allende has joined hundreds of other governments, regions, municipalities, and NGOs pledging to reduce and cancel out greenhouse gas emissions as soon as possible, utilizing not only reductions in fossil fuel emissions through energy conservation and the conversion to renewable energy, but also utilizing the massive sequestration potential of native agave plants and companion nitrogen-fixing trees to draw down and store large amounts of excess atmospheric carbon in soils, trees, and plants through the enhanced photosynthesis of regenerative farming and land use practices.

The deployment of the agave agroforestry system on approximately 45,734 hectares (114,335 acres) over 10 years can enable the municipality of San Miguel de Allende to sequester and store enough atmospheric carbon to cancel out all of its projected greenhouse gas emissions over 10 years and thereby fulfill its pledge in the "4 per 1000" Initiative to reach zero net greenhouse gas emissions by 2033.

There are 2,400 *municipalidades* in Mexico, including 1,000 that already have substantial amounts of agave growing wild and/or being semi-cultivated. A number of these *municipalidades* are already cultivating agave and harvesting the *piñas* for mescal. For example, in the watershed of Tambula Picachos in the municipality of San Miguel there are 39,022 hectares of rural land (mainly *ejido* land) in need of restoration (93.4 percent show signs of erosion, 53 percent with compacted soil). Deploying the agave agroforestry system on most of this degraded land would be enough to cancel out most current annual emissions in the municipality of San Miguel.

The gross economic value of growing agave on 50 percent of these 39,022 hectares (including silage, *piñas*, and *hijuelos*) at $20,000 per year (averaged out over 10 years) would amount to around $390 million per year, a tremendous boost to the economy. In comparison, San Miguel de Allende, one of the top tourist destinations in Mexico (with 1.3 million visitors annually), brings in $1 billion a year from tourism, its top revenue generator.

With the agave agroforestry system, several million small dryland farmers in Mexico and many thousands of private landowners can restore their degraded lands and revitalize rural landscapes across the nation.

Several hundred million farmers in the world's semiarid and arid lands, 40 percent of the world's terrain, can do the same.

In chapter 7, we will look at how the agave agroforestry system, along with other game-changing best practices of organic and regenerative farming and forestry, can be financed and scaled up in Mexico and in nations across the world.

CHAPTER SEVEN

Scaling Up

Ronnie and I saw an urgent need to scale up agroecological, regenerative, and organic systems to regenerate our ecosystems, climate, and communities. We have been actively involved in organic agriculture in various ways for decades. I have been and continue to be an organic farmer since the early 1970s.

We spent several years researching this, actively looking at the markets, the standards, and verification systems. The more we went into this, the more concerned we were about the credibility of these carbon offset schemes. Indeed, global media has recently been highly critical of carbon offsets, and journalists and scientists have found that many of these systems are fraudulent.

Blood Carbon

Countries and industries fund dubious carbon offset schemes to justify increasing their fossil fuel emissions. The worst of these are called "blood carbon" due to the way they are evicting Indigenous owners from their lands. It is a repeat of colonialism, as they are stealing the land for these schemes.

A distressing example is the Framework of Collaboration between the United Arab Emirates' firm Blue Carbon and the Republic of Kenya for developing REDD+ projects and the development of carbon credits for millions of acres, signed on October 25, 2023. (REDD is the acronym for the United Nations agreement for Reducing Emissions from Deforestation and Forest Degradation in Developing Countries. REDD+ includes sustainable forest management and conservation, and the enhancement of forest carbon stocks.)

Blue Carbon also signed memoranda of understanding with Liberia, Zambia, Tanzania, and Pakistan, and announced a partnership with the Republic of Zimbabwe for developing forest projects on 18.75 million acres (7.5 million hectares) in exchange for $1.5 billion in carbon credits.

Before Kenya signed the agreement with Blue Carbon, Kenyan President William Ruto ordered security agencies to remove the Ogiek people from the Mau Forest, the largest forest in the country, in order to "drive Kenya's climate change action." Forest rangers would evict the "illegal settlers," who were causing the "wanton destruction of forests." He stated that protecting Mau forest resources was necessary to fight the climate crisis. In fact, the Ogiek, which translates to "caretaker of all plants and wild animals," have lived in the Mau forest since time immemorial, successfully managing its complex ecosystem, with around 35,000 to 45,000 people coexisting on its 1 million acres (400,000 hectares).

The Ogiek people have been subjected to brutal evictions and land grabs since the days of British colonial rule, and since independence, Kenyan authorities have carried out many violent and brutal evictions, destroying homes and property and killing people who attempted to protect their land. Corrupt officials gave the land to their political cronies, who logged the forest for profit. In 2017, the African Court on Human and Peoples' Rights (ACHPR) ruled that the government had violated the Ogiek's land rights and explicitly recognized their crucial role in conserving and protecting the Mau

Forest. Nevertheless, the Kenyan government ignored the ACHPR ruling and evicted additional Ogiek communities from the Mau Forest. Filing with the ACHPR a second time, the Ogiek won a ruling in 2022, stating that the government owed the Ogiek reparations for failing to comply with the 2017 ruling.

As I write this in November 2023, however, the Ogiek are being evicted yet again on the basis that they are "illegal settlers," causing the "wanton destruction of forests." The Ogiek have responded by saying that they are the forest owners and protectors. The damage to the forest has been caused by the corrupt allocation of land to loggers and others who cut down the trees to sell timber or charcoal.

Rangers from the Forestry and Wildlife Services, in collaboration with the police, are now openly defying the court ruling that asserts the Ogiek's ownership rights to the land, and have illegally evicted up to 700 Ogiek people, destroying their homes by either dismantling or burning them. What's happening to the Ogiek people is a massive human rights crime, and just one among many such examples of Indigenous communities who are confronting the theft of their land under the pretense of climate change mitigation.[1]

Rewarding Ecosystem Services

Rather than evicting Indigenous peoples like the Ogiek, these traditional land managers should be financially rewarded for their valuable ecosystem services, including the maintenance and regeneration of natural and agricultural ecosystems and soils, and carbon dioxide removal. A 2023 study published in *Nature Climate Change* compared 314 forest sites managed by local people in 15 tropical countries in Africa, Asia, and Latin America, and found that forests managed by Indigenous and local communities have *improved* outcomes for carbon, biodiversity, and livelihoods.

Examining the relationships between carbon sequestered in aboveground woody biomass, tree species numbers (a proxy for biodiversity), and forest livelihoods, the researchers found that the presence of formal community management and local participation in rulemaking were consistent predictors of multiple positive outcomes. Their findings from various global contexts show that empowered local forest governance supports multiple forest restoration objectives. They stated: "Our most striking finding is that empowered local governance—in the form of formal community forest management organizations and local participation in rule-making—is a key predictor of multiple positive outcomes. This finding aligns with a well-established body of research on how local actors possess a comparative advantage for coordinating local governance functions."[2]

These findings underscore the importance of giving rural and Indigenous communities formal, legally recognized opportunities to engage in local management practices. This critical step is needed to advance multiple human and environmental benefits in forested landscapes worldwide.

The Perversity of Additionality

The UN requires that all payments for carbon dioxide drawdown must result from additional CO_2 being removed through a management change. Existing forests or farming practices that are removing CO_2 do not receive funding since that CO_2 is not regarded as additional sequestration. It is assumed that these systems would continue via business as usual and not contribute to the additional CO_2 that must be removed to reduce atmospheric levels. This is called additionality.

For example, an area that has been cleared and replanted with trees will receive payment for the CO_2 sequestered by the trees. An existing forest that has been carefully managed will not receive any payments for the CO_2 it has already sequestered and the CO_2 it will sequester. In other words, additionality penalizes landholders who have done the right thing by conserving existing biodiversity or increasing their soil carbon through good management instead of rewarding them for avoided emissions and ecosystem services.

The UN's application of additionality is stealing from the land managers with systems that continuously remove CO_2 from the atmosphere. Without these natural agroecosystems, greenhouse gas levels would be much higher, and biodiversity loss and climate change would be far worse. The value of these systems needs to be recognized, and the traditional owners and managers of these systems need to be financially rewarded.

Paying for Ecosystem Services

Instead of trading carbon as a commodity on financial markets, polluters should be required to remunerate the people who are actually doing the work of regenerating the climate and environment. Currently, these services are being provided for free. Landholders who are doing the right thing are being exploited.

The world must recognize that climate change, biodiversity loss, rainfall, water quality, and other metrics would be far worse without these services. It is time to pay for the actual value of these services instead of taking them for granted. This is real climate change and environmental justice. Paying for ecosystem services such as regenerating and conserving biodiversity, increasing carbon dioxide removal, and improving soil health and water quality will incentivize a change in land management from the current trend of widespread degeneration to widespread regeneration.

Trillions of dollars have gone into various United Nations organizations, carbon-credit schemes, and industrial-scale carbon capture and storage proposals over the decades, with very little to show except for bulging bureaucracies with highly paid staff who produce reports and fund projects that funnel money through corrupt governments and billionaire cartels. These actors have failed to reduce greenhouse gas emissions, clearcutting, biodiversity loss, wars and conflicts, droughts, fires, and floods, and the malnourishment and starvation of billions living in crushing poverty.

Instead of continuing to waste trillions of dollars, financing must be redirected to compensate those providing the ecosystem services to regenerate our climate, environments, food and farming systems, and communities. This needs to be done with a new program that is not connected to the current corrupt and ineffective global system. These ecosystem services are certainly worth far more than clearing land for large-scale toxic monocultures of GMO soy, palm oil, beef, and the extraction of timber and minerals.

In previous chapters we have outlined the multiple benefits of regenerative practices. These benefits include creating rainfall, improving regional cooling, reversing vapor pressure deficits, and increasing soil organic matter. These benefits have tangible values that need to be paid for rather than just taken for free, neglected, and therefore regarded as worthless. Paying for them will give them real value.

Equally important is the need to regenerate our food and farming systems away from toxic monocultures. The previous chapters outlined how these systems destroy soil organic matter and reduce biodiversity, destroying soil's capacity to capture and retain water. This leads to vapor pressure deficits that heat up the land, causing both droughts and torrential flooding rains.

Scaling up diverse perennial agroforestry systems is critical. The previous chapters offered many examples of agroecological systems that are based on perennial trees, shrubs, and herbs combined with annual crops. Permanent covers of photosynthesizing plants are essential to draw down CO_2, increase soil organic matter, increase transpiration to cool the region, and reduce vapor pressure deficits. Scaling up tree and forest regeneration is the fastest way to cool the planet. The world is currently about to become 2.7°F (1.5°C) warmer; research shows that adequate tree cover would increase transpiration and provide regional cooling to more than compensate for this.

Because CO_2 persists for about 1,000 years in the atmosphere, it will take centuries for the atmospheric heat to dissipate unless it is actively removed to 1750s, pre-Industrial Revolution levels. Furthermore, most excess heat is trapped in the oceans, making the situation even more dire. There are agroforestry systems that can be scaled up in all climates, however. The agave agroforestry system can be scaled up in the arid and semiarid regions of all six arable continents. These are some of the most degraded regions of the planet due to degenerative grazing and farming practices, but nevertheless support hundreds of millions of people who live there and rely on these lands.

A Call to Action

In chapter 2, I explained that CO_2 emissions have increased by about 2.87 ppm annually since the 2015 Paris Agreement and that this rate continues to grow. We need to draw down

28 gigatons of CO_2-eq per year to stop the increase of CO_2 in the atmosphere and achieve net zero emissions. After that we then need to remove even more CO_2 to prevent the temperature rise from exceeding 2.7°F (1.5°C). We must reach net zero and achieve net negative (reverse) emissions as soon as possible to avoid runaway global warming, wholesale biodiversity collapse, climate catastrophe, endless poverty-driven conflict, forced migration, and wars. Even if the world transitioned to 100 percent renewable energy tomorrow, it would not stop the rise in temperature and sea levels since CO_2 persists for over 1,000 years in the atmosphere and the oceans' heat will continue adversely affecting the climate until it slowly dissipates.[3]

We are in a climate emergency now. We must reduce all types of greenhouse gas emissions, speed up the transition to non-degenerative forms of renewable energy, stop clearing forests, regenerate ecosystems, and make a great effort to use nature-based systems to sequester CO_2 from the atmosphere. Besides making the transition to environmentally appropriate forms of renewable energy and implementing long-overdue energy conservation measures, the global organic, agroecological, and regenerative community needs to draw down and sequester, aboveground and belowground, more than 28 gigatons of CO_2-eq per year in our soils, plants, and trees, as soon as possible.

The planet's current and potential terrestrial carbon sink and biodiversity habitat comprises 4 billion acres of croplands, 8 billion acres of pasturelands and rangelands, and 10 billion acres of forest. Besides planting 1 trillion new trees and preserving and improving our 10 billion acres of forest lands, we need to bring 1 billion acres of cropland and rangeland (out of the total of 12 billion) under organic, agroecological, and regenerative management.

In chapter 3, we showed how scaling up 5 to 10 percent of agricultural lands with a few selected best-practice regenerative systems could draw down more than the current emissions—so that the world would go into negative emissions and reverse climate change. If we can do this, we will not only be able to reverse global warming; we can stop the great extinction of biodiversity currently underway, begin to reverse the global water crisis, qualitatively improve public health and nutrition, and eliminate poverty among the 3 billion farmers, farmworkers, and rural villagers on the planet.

A Framework for Rewarding Ecosystem Services

As you now know, an enormous amount of money is spent on mitigating climate change, yet the problems only worsen. The latest analysis by Bloomberg suggests that global environment, social, and governance (ESG) assets will exceed $53 trillion by 2025. Asset managers use ESG as a major criterion in investing their funds in corporations. ESG will represent over a third of the $140.5 trillion in projected total assets under management. Climate change and environmental management are major factors in investment decisions.

According to the Boston Consulting Group, in 2021 the voluntary carbon market grew rapidly, reaching $2 billion, four times its value in 2020. The group predicts that the market will reach between $10 billion and $40 billion by 2030. Billions of dollars have been spent on carbon capture and storage technologies, most of which are net emitters of greenhouse gases. Governments are spending billions on subsidizing the fossil fuel industry and industrial-scale renewable energy schemes.

The carbon credits, environmental asset derivatives, and ESG programs primarily benefit traders, scheme owners, government bureaucrats, and consultants. Those of us who work on the front line and visit the communities see few tangible benefits. The people who manage the ecosystems rarely see much of the funds. Most of it is spent before it gets to them.

In previous chapters we have presented strong evidence that regenerating ecosystems and farmlands through growing more plants and increasing soil organic matter can sequester all the current emissions and cool the planet. We need just a percentage of the trillions of dollars that are currently being misused to be redirected into regenerative systems.

Incentive to Change

Currently, farmers are only paid for their yields, not their ecosystem services. The current system favors farmers who can produce the cheapest commodities using economies of scale. This rewards a race to the bottom—the owners of CAFOs and large-scale monocultures with their damaging environmental, health, and social effects reap the benefits. The true cost of damaging the climate, biodiversity, human health, and communities will be paid later—by future generations. These systems are stealing from our children, grandchildren, and those yet to be born.

The current system treats carbon as a tradable commodity—with arbitrary and, in many cases, illogical rules around permanency and additionality. Climate change meetings and academic and political discussions are tantamount to walking in endless circles. Participants fiddle over academic and political disagreements while Rome burns.

Instead of treating carbon as a tradable commodity that rewards financial markets, commodity exchanges, traders, scheme owners, and consultants, the approach that Ronnie and I discussed before he passed away involves the payment of a fee for service. This would change the focus. When we pay for a plumber, dentist, lawyer, doctor, mechanic, or accountant, we aren't paying them for specific commodities. We pay them for the results of their service.

Paying for the services of removing CO_2 and regenerating ecosystems, such as biodiversity and climate, would result in a massive financial incentive for changes in land management practices. Instead of a race to the bottom to produce commodities for profit, there would be an incentive for regenerating ecosystems and food and farming production systems.

The Agroecological, Regenerative, and Organic Ecosystem Services (AROES) Framework

Ronnie and I developed the initial draft of the Agroecological, Regenerative, and Organic Ecosystem Services (AROES) framework to pay landholders for ecosystem services. As the draft took shape, we sought input from the Regeneration International management team, including our continental coordinators: Ercilia Sahores for Latin America, Precious Phiri for Africa, Oliver Gardiner for Eurasia, Mercedes Lopez from Vía Orgánica, and our finance manager Rose Welch.

Regeneration International is currently setting up and managing AROES as a registry and secretariat to validate and coordinate all the primary services and payments. It is based on payments to farmers, traditional owners, and land managers for:

- Conserving and/or regenerating biodiversity
- Removing carbon dioxide through aboveground biodiversity and/or soil organic carbon to reverse climate change

- Improving gender equity
- Improving fairness in labor, production, and marketing

Regeneration International and our partners can achieve multiple objectives with adequate funding through public education, market demand, farmer-to-farmer training, grass-roots lobbying, and policy reform. This registry will develop, clarify, and channel financial incentives and investments into the ecological goods and services marketplace.

We will use organic certification systems to verify ecological outcomes tied to management practices to develop a system that pays producers for ecosystem services. It will connect to purchasers of those services. It will identify and work with ethical companies that can finance the scaling up of these best practices and improve market access and infrastructure development for producers, communities, and regions.

Monitoring, Reporting, and Verification (MRV)

Emerging stories of fraudulent carbon offset schemes have damaged the credibility of the carbon offset markets. Companies now demand a reputable monitoring, reporting, and verification (MRV) framework as a top credit purchasing criterion. (Over 90 percent of buyers consider MRV a significant factor in credit purchase decisions.) The credits must have a verification system to clearly show their impact and that can be used to defend their credibility against the ever-growing claims of greenwashing.

Organic certification systems are the world's oldest, most reliable, and most trusted agricultural verification systems. They are based on internationally accepted best practices. Regeneration International will use credible organic certifiers and Participatory Guarantee System (PGS) verification systems to verify our ecosystem projects, combined with our own purpose-developed AROES standard to suit the precise purposes of regeneration. Certified operators will also have the option of being certified to national organic standards for market access, including major standards such as the USDA National Organic Program (NOP), the EU regulation, the Japanese Agricultural Standards (JAS), and so on.

The AROES standard is short, simple, and straightforward rather than lengthy or complex. It is designed to be easy for farmers and landholders to use. Most farmers in the developing world have limited education and cannot understand complicated certification standards. The standard has clear prohibitions of degenerative practices and inputs. These include animal cruelty, CAFOs, hydroponics, GMOs, chemical fertilizers, synthetic pesticides, and damaging tillage. It uses guidance rather than mandated practices so that farmers can select the most appropriate to make their decisions. The success of these practices will be measured using evidence-based results as part of the MRV processes.

Teams of trained experts will conduct measurements and provide technical expertise and objective results. These measurement systems will be simple and practical and will not employ highly complex scientific methodologies. New technologies such as Light Detection and Ranging (LiDAR) are being used to improve accuracy, save time, and reduce MRV costs.

The use of proxies is a key component to reduce costs and workloads. For example, soil organic matter levels are a good proxy for soil health and soil carbon sequestration. It is easily measured and can be used to determine how well a farmer is regenerating his or her soils. Tree and plant diversity and bird calls are good proxies for biodiversity. For example, increases in the number of plant species are a sign of healthy ecosystems and biodiversity, as are increases in bird calls. These proxies can be used as the basis for ecosystem services payments.

Regeneration International will also approve and accredit the certifiers and certification systems to ensure every project's veracity. Consequently, these projects will be as robust and credible as possible, with the highest verification and fraud detection levels.

A Worldwide Network

As the umbrella of a worldwide network of more than 570 partners, Regeneration International will multiply the number of certified farmers and acreage by using financial and agronomic incentives to encourage and motivate producers to adopt the best organic, regenerative, and agroecological practices. Our network building is designed to be scalable. We expect it to multiply, especially in the Global South, as farmers and land managers learn the benefits of adopting agroecological, regenerative, and organic best management practices, verifying them, and then getting paid for them. The first pilot projects have started and will become catalysts for change in their communities.

AROES will not pay for carbon as a commodity since it belongs in the soil or vegetation. Landholders will be paid for their verified management services of removing it from the atmosphere—as negative emissions. The corporations emitting greenhouse gases will pay them for doing this valuable service and can use the verified carbon dioxide removal numbers to reduce their carbon footprints. That is natural climate justice. The funds will go to the people who deserve it—the landholders regenerating soils and ecosystems—rather than the UN, nation-states, or private-sector carbon credit systems that reward consultants, bureaucracy, and corruption.

AROES will replace the widely criticized carbon credits and offsets with a new system that pays and incentivizes producers to provide multiple ecosystem services. These will include paying for verifiable CO_2 removal and avoidance services that the UN states are essential to ensure the world does not exceed 2.7°F (1.5°C).

This project is building a registry that will expand the incentives to include soil health, water, biodiversity, and economic justice, offering significant financial rewards for producers. In this way, we can motivate a critical mass of the world's farmers and livestock managers to take their practices to the next level (organic and regenerative based on the science and practices of agroecology), making them eligible for payments for a range of ESG ecosystem services.

We only need to transition a small proportion of agricultural production to best-practice agroecological, regenerative, and organic systems to:

- Remove enough CO_2 to draw down more than the current emissions
- Regenerate ecosystems that will cool their regions and reduce droughts and heatwaves
- Increase soil organic matter to capture and retain rainfall to rehydrate the landscape and improve crop yields
- Produce higher yields of healthy food with no toxic chemicals
- Regenerate our communities, health, climate, and environment

There is no need to close farms, kill livestock, or destroy communities or the environment. These are shovel-ready solutions!

The proliferation of devastating storms, droughts, floods, fires, wars, biodiversity loss, mass extinctions, health epidemics, and food insecurity is certainly not good news. But good news does exist: We can turn things around by regenerating our planet. We know how to do it—it isn't rocket science. Many of us are doing this now, and we need many others to join us. Together, we can give ourselves, our children, and all the living species we share our planet with a great future.

Ronnie Cummins wrote: "Never underestimate the power of one individual: yourself. But please understand, at the same time, that what we do as individuals will never be enough. We've got to get organized, and we've got to help others in our region, our nation, and everywhere build a mighty green regeneration movement. The time to begin is now."

APPENDIX

Calculations

In chapter 3, I reported my "back-of-the-envelope" calculations for achieving negative emissions with a sample of regenerative agriculture systems. These calculations are not intended as scientific proof; rather, these types of back-of-the-envelope analyses are intended to help conceptualize the potential of a strategy or methodology when testing a hypothesis. They are a starting point, not an ending point.

By the same token, exact calculations of anthropogenic greenhouse gas emissions need to be taken with a grain of salt because governments and organizations use different methodologies to calculate them from numerous sources. When variations of estimated figures are published, their credibility must always be questioned.

For simplicity, I have chosen to use one objective measure: the parts per million (ppm) of CO_2 recorded at the NOAA Mauna Loa Observatory in Hawaii. According to their data, CO_2 emissions have increased by about 2.87 ppm per year since the 2015 Paris Agreement. Using an average of eight years of data is more reliable than selecting one year due to annual variability.

There are ongoing scientific debates about carbon dioxide's exact contribution to trapping infrared energy compared to the other anthropogenic greenhouse gases. I have read many studies with variable results.

The study that has the most comprehensive datasets and solid methodology states that CO_2 is responsible for 20 percent of the total energy increase, with the other anthropogenic greenhouse gases responsible for 5 percent and water vapor and clouds responsible for 75 percent.[1]

CO_2-eq combines carbon dioxide and other anthropogenic greenhouse gases such as methane, nitrous oxide, and halocarbons. Their warming potentials are expressed in units equivalent to CO_2. The combination of all these greenhouse gases is used to better understand the levels of gases contributing to global warming. CO_2 accounts for 80 percent of all human-produced greenhouse gases, while the others account for 20 percent.

A reasonable assumption is that 2.87 ppm of CO_2 is 80 percent of carbon dioxide equivalents (CO_2-eq), which would give a figure of 3.6 ppm CO_2-eq.

Using the accepted formula that 1 ppm CO_2 = 7.76 gigatons (Gt) CO_2, we will need to sequester 28 Gt of CO_2-eq per year to stop further increases of CO_2 in the atmosphere. However, we'd have to draw down even more than this (with negative emissions) to reduce CO_2 and limit temperature rise to 1.5°C (2.7°F).

For the five examples I have chosen—the agave agroforestry system, adaptive multi-paddock (AMP) grazing, pasture cropping, biologically enhanced agricultural management (BEAM), and "no kill, no till"—I am basing land use figures from the United Nations Food and Agriculture Organization (UN FAO). The UN FAO has estimated that the total amount of land used worldwide to produce food and fiber is 4,911,622,700 hectares (12,279,056,750 acres). This is divided into:

- Arable cropland: 1,396,374,300 hectares (3,490,935,750 acres)
- Permanent pastures: 3,358,567,600 hectares (8,396,419,000 acres)
- Permanent crops: 153,733,800 hectares (384,334,500 acres)

The Agave Agroforestry System

Research by Dr. Mike Howard shows that the agave agroforestry system can sequester 8.7 metric tons of carbon dioxide per hectare per year (8,700 pounds per acre).

This is without counting belowground soil organic matter sequestration nor the amount of carbon sequestered by the companion trees. Extrapolated globally across the 3.3 billion hectares of permanent pastures, the agave agroforestry system could sequester 28.7 Gt of CO_2 annually. This is slightly more than the 28 Gt of CO_2-eq per year that is currently emitted, and its sequestration would start to reverse climate change and regenerate the climate. If this system were deployed on 10 percent of the permanent pastures, it could sequester 2.8 Gt of CO_2 per year.

Regenerative and Adaptive Multi-Paddock (AMP) Grazing

Megan Machmuller's research demonstrates that regenerative grazing in the United States can sequester 29.36 metric tons of CO_2 per hectare per year (29,360 pounds per acre).[2]

If these regenerative grazing practices were implemented on all of the world's permanent pastures, they would sequester 98.6 Gt of CO_2 per year. (29.36 tons of CO_2 per hectare per year × 3,358,567,600 hectares = 98,607,544,736 tons of CO_2 per hectare per year.) If this system were deployed on 10 percent of the world's grazing lands, they could sequester 9.86 Gt of CO_2 per year.

Pasture Cropping

Neils Olsen's pasture cropping system sequestered 13 metric tons of CO_2 per hectare per year (13,000 pounds per acre). If this were applied to permanent pastures and arable cropland, it would sequester 63.8 Gt of CO_2 per year. (4,754,941,900 hectares × 13 tons CO_2 per hectare per year = 61,814,244,700 tons of CO_2 per hectare per year.)

If this system were deployed on 10 percent of all permanent pastures and arable/croplands, it could sequester 6.18 Gt of CO_2 per year.

Biologically Enhanced Agricultural Management (BEAM)

BEAM sequestered 37.7 metric tons of CO_2 per hectare per year (37,700 pounds per acre).

If BEAM were extrapolated globally across agricultural lands, it would sequester 185 Gt of CO_2 per year (37.7 tons of CO_2 per hectare per year × 4,911,622,700 hectares = 185,168,175,790 tons of CO_2 per hectare per year.)

If this system were deployed on 5 percent of agricultural lands, it could sequester 9.2 Gt of CO_2 per year.

No Kill, No Till

Paul and Elizabeth Kaiser of Singing Frog Farm have managed to increase their soil organic matter from 2.4 percent to an optimal 7 to 8 percent in just 6 years, an average increase of about .75 percentage points per year. According to Dr. Christine Jones, "An increase of 1 percent in the level of soil carbon in the 0–30 cm soil profile equates to sequestration of 154 tons CO_2/ha with an average bulk density of 1.4 g/cm3."[3] It follows that .75 percent organic matter = 115.5 metric tons of CO_2 per hectare (115,500 pounds an acre per year).

This system can be used on arable and permanent croplands for a total of 1,550,108,100 hectares (3,875,270,250 acres). Extrapolated globally across arable and permanent croplands it would sequester 179 Gt of CO_2 per year (1,550,108,100 hectares × 115.5 metric tons of CO_2 per hectare = 179,037,485,550 metric tons.) If this system were deployed on 5 percent of arable and permanent croplands, it could sequester 8.9 Gt of CO_2 per year.

Conversion Formulas

1°C = 1.8°F is the standard formula for converting Celsius to Fahrenheit.

There are 2.471054 acres to a hectare. I have rounded this off to 2.5 acres to the hectare, a standard practice.

Similarly, I have rounded off pounds per acre to kilograms per hectare.

Metric tons and US tons are approximate.

A gigaton (Gt) is one billion tons.

Notes

Chapter 1. Epiphany in the Desert

1. Most technical papers use the term soil organic *carbon* (SOC) when measuring the amount of carbon in soils. By contrast, the term soil organic *matter* (SOM) is largely used in farming systems, especially as a proxy for soil health and benefits. As a general rule, the higher the levels of SOM, the greater the benefits in terms of soil fertility, water capture and retention, friability, and the soil microbiome. Because SOM is more than just carbon, there is a ratio in which one unit of SOC equals 1.72 units of SOM. Note, however, that for the sake of this book, in most cases, the terms SOC and SOM are interchangeable.

Chapter 2. Climate Science, Skepticism, and "Solutions"

1. "Global Land Outlook: Land Restoration for Recovery and Resilience, 2nd ed.," UN Convention to Combat Desertification, April 27, 2022, https://www.unccd.int/sites/default/files/2022-04/UNCCD_GLO2_low-res_2.pdf.
2. G. A. Schmidt et al., "Attribution of the Present-Day Total Greenhouse Effect," *Journal of Geophysical Research* 115 (October 2010): D20106, https://doi.org/10.1029/2010JD014287.
3. Rebecca Lindsey and LuAnn Dahlman, "Climate Change: Ocean Heat Content," NOAA, September 6, 2023, https://www.climate.gov/news-features/understanding-climate/climate-change-ocean-heat-content.

4. Lindsey and Dahlman, "Climate Change"; Zhi Li, Matthew H. England, and Sjoerd Groeskamp, "Recent acceleration in global ocean heat accumulation by mode and intermediate waters," *Nature Communications* 14 (October 2023): 6888, https://doi.org/10.1038/s41467-023-42468-z.
5. James P. Kossin et al., "Global Increase in Major Tropical Cyclone Exceedance Probability Over the Past Four Decades," *Proceedings of the National Academy of Sciences* 117, no. 22 (June 2020): 11975–80, https://doi.org/10.1073/pnas.1920849117.
6. Tom Knutson, "Global Warming and Hurricanes: An Overview of Current Research Results," *Geophysical Fluid Dynamics Laboratory*, accessed November 8, 2023, https://www.gfdl.noaa.gov/global-warming-and-hurricanes.
7. Hannah Ritchie, "Deforestation and Forest Loss," *Our World in Data*, February 4, 2021, https://ourworldindata.org/deforestation.
8. Yann Arthus-Bertrand, "On Water," *European Investment Bank*, March 22, 2018, https://www.eib.org/en/essays/on-water.
9. Ritchie, "Deforestation and Forest Loss."
10. IPCC 2019a, 367.
11. Yan Li et al., "Local Cooling and Warming Effects of Forests Based on Satellite Observations," *Nature Communications* 6 (March 2015): 6603, https://doi.org/10.1038/ncomms7603.
12. Edward W. Butt et al., "Amazon Deforestation Causes Strong Regional Warming," *PNAS* 120, no. 45 (November 2023): e2309123120, https://doi.org/10.1073/pnas.2309123120.
13. Butt et al., "Amazon Deforestation Causes Strong Regional Warming."

14. Almudena García-García et al., "Soil Heat Extremes Can Outpace Air Temperature Extremes," *Nature Climate Change* 13 (September 2023):1237–41, https://doi.org/10.1038/s41558-023-01812-3.
15. Wenping Yuan et al., "Increased Atmospheric Vapor Pressure Deficit Reduces Global Vegetation Growth," *Science Advances* 5, no. 8 (August 2019), https://doi.org/10.1126/sciadv.aax1396.
16. M. Fujita et al., "Precipitation Changes in a Climate With 2-K Surface Warming From Large Ensemble Simulations Using 60-Km Global and 20-Km Regional Atmospheric Models," *Geophysical Research Letters* 46, no. 1 (January 2019): 435–42, https://doi.org/10.1029/2018GL079885.
17. Yuan et al., "Increased Atmospheric Vapor Pressure."
18. Yitao Li et al., "Biophysical Impacts of Earth Greening Can Substantially Mitigate Regional Land Surface Temperature Warming," *Nature Communications* 14 (January 2023): 121, https://doi.org/10.1038/s41467-023-35799-4.
19. Yuan et al., "Increased Atmospheric Vapor Pressure."
20. Kimberly A. Novick et al., "The Increasing Importance of Atmospheric Demand for Ecosystem Water and Carbon Fluxes," *Nature Climate Change* 6 (September 2016): 1023–27, https://doi.org/10.1038/nclimate3114.
21. Richard Seager et al., "Climatology, Variability, and Trends in the U.S. Vapor Pressure Deficit, an Important Fire-Related Meteorological Quantity," *Journal of Applied Meteorology and Climatology* 54, no. 6 (June 2015): 1121–41, https://doi.org/10.1175/JAMC-D-14-0321.1.
22. "Trends in Atmospheric Carbon Dioxide," Global Monitoring Laboratory, Earth System Research Laboratories, https://gml.noaa.gov/ccgg/trends/weekly.html.

23. Ian Tiseo, "Average Carbon Dioxide (CO_2) Levels in the Atmosphere Worldwide From 1959 to 2023," Statista, accessed August 27, 2022, https://www.statista.com/statistics/1091926/atmospheric-concentration-of-CO2-historic.
24. "The Paris Agreement," United Nations Framework Convention on Climate Change, 2015, https://unfccc.int/process-and-meetings/the-paris-agreement/the-paris-agreement.
25. Working Group III to the Fifth Assessment Report of the Intergovernmental Panel on Climate Change [O. Edenhofer et al., ed.], "Summary for Policymakers," in *Climate Change 2014: Mitigation of Climate Change* (Cambridge, UK and New York, NY, USA: Cambridge University Press, 2014), https://www.ipcc.ch/site/assets/uploads/2018/02/ipcc_wg3_ar5_full.pdf, 4-30.
26. IPCC Working Group III, "Summary for Policymakers," 2014.
27. IPCC Working Group III, "Summary for Policymakers," 2014.
28. V. Masson-Delmote et al., ed., "Summary for Policymakers," in *Global Warming of 1.5°C. An IPCC Special Report on the Impacts of Global Warming of 1.5°C Above Pre-Industrial Levels and Related Global Greenhouse Gas Emission Pathways, in the Context of Strengthening the Global Response to the Threat of Climate Change, Sustainable Development, and Efforts to Eradicate Poverty* (Cambridge, UK and New York, NY, USA: Cambridge University Press, 2018), 3–24, https://www.ipcc.ch/sr15/chapter/spm.
29. Working Group III to the Sixth Assessment Report of the Intergovernmental Panel on Climate Change [P. R. Shukla et al., ed.], "Summary for Policymakers,"

in *Climate Change 2022: Mitigation of Climate Change* (Cambridge, UK and New York, NY, USA: Cambridge University Press, 2022), accessed Aug 27, 2022, https://www.ipcc.ch/report/ar6/wg3/downloads/report/IPCC_AR6_WGIII_SPM.pdf.

30. Bruce Robertson and Milad Mousavian, "The Carbon Capture Crux: Lessons Learned," Institute for Energy Economics and Financial Analysis, September 2020, https://ieefa.org/resources/carbon-capture-remains-risky-investment-achieving-decarbonisation.

Chapter 3. The Promise and Potential of Regenerative Agriculture

1. Calculations depend on the boundaries and methodologies used to determine the emissions.
2. FAO 2010, Global Livestock Environmental Assessment Model (GLEAM), accessed September 1, 2022, https://www.fao.org/gleam/en. See appendix for how I arrived at calculations.
3. André Leu, "Our Global Regeneration Revolution: Organic 3.0 to Regenerative and Organic Agriculture," Regeneration International, July 7, 2021, https://regenerationinternational.org/2021/07/12/our-global-regeneration-revolution-organic-3-0-to-regenerative-and-organic-agriculture.
4. "Our Story," Rodale Institute, https://rodaleinstitute.org/about-us/mission-and-history.
5. Miguel A. Altieri, "Agroecology: The Science of Natural Resource Management for Poor Farmers in Marginal Environments," *Agriculture, Ecosystems & Environment* 93, no. 1–3 (December 2002): 1–24, https://doi.org/10.1016/S0167-8809(02)00085-3.
6. Inga Alley Cropping, https://www.ingafoundation.org/alley-cropping.

7. W. R. Teague et al., "Grazing Management Impacts on Vegetation, Soil Biota and Soil Chemical, Physical and Hydrological Properties in Tall Grass Prairie," *Agriculture, Ecosystems & Environment* 141, no. 3–4 (May 2011): 310–22, https://doi.org/10.1016/j.agee.2011.03.009.
8. W. R. Teague et al., "The Role of Ruminants in Reducing Agriculture's Carbon Footprint in North America," *Journal of Soil and Water Conservation* 71, no. 2 (March/April 2016): 156–64, https://doi.org/10.2489/jswc.71.2.156.
9. Tong Wang et al., "GHG Mitigation Potential of Different Grazing Strategies in the United States Southern Great Plains," *Sustainability* 7, no. 10 (September 2015): 13500–521, https://doi.org/10.3390/su71013500.
10. Megan B. Machmuller et al., "Emerging Land Use Practices Rapidly Increase Soil Organic Matter," *Nature* Communications 6 (April 2015): 6995, https://doi.org/10.1038/ncomms7995.
11. André Leu, "Commentary V: Mitigating Climate Change with Soil Organic Material in Organic Production Systems," in "Chapter 1: Key Development Challenges of a Fundamental Transformation of Agriculture," in *Trade and Environment Review 2013: Wake Up Before It Is Too Late*, ed. Ulrich Hoffmann (UNCTAD, 2013), 22.
12. Teague et al., "The Role of Ruminants."
13. Wang et al., "GHG Mitigation Potential."
14. "Pasture Cropping," Winona, accessed September 4, 2022, https://winona.net.au/pasture-cropping.
15. Soilkee, "World First at the Soilkee Farm, with First Carbon Credits Issued to a Soil Carbon Project Under the Emissions Reduction Fund and the Paris Agreement," Soilkee media release, March 14, 2019, https://soilkee.com.au/gallery/20190314-soilkee%20media%20release%20.pdf.

16. "Emissions Reduction Case Studies," Australian Government Clean Energy Regulator, accessed September 4, 2022, https://www.cleanenergyregulator.gov.au/Infohub/case-studiesemissions-reduction-fund-case-studies#Soil-carbon.
17. Cindy E. Prescott et al., "Managing Plant Surplus Carbon to Generate Soil Organic Matter in Regenerative Agriculture," *Journal of Soil and Water Conservation* 76, no. 6 (November 2021): 99A–104A, https://doi.org/10.2489/jswc.2021.0920A.
18. David Johnson, Joe Ellington, and Wesley Eaton, "Development of Soil Microbial Communities for Promoting Sustainability in Agriculture and a Global Carbon Fix," *PeerJ PrePrints* (January 13, 2015), https://doi.org/10.7287/peerj.preprints.789v1/supp-1.
19. Personal communication with the author.
20. Neha Begill, Axel Don, and Christopher Poeplau, "No Detectable Upper Limit of Mineral-Associated Organic Carbon in Temperate Agricultural Soils," *Global Change Biology* 29, no. 16 (June 2023): 4662–69, https://doi.org/10.1111/gcb.16804.
21. André Leu, *Growing Life: Regenerating Farming and Ranching* (Greeley, CO: Acres USA, 2021); Dayakar V. Badri and Jorge M. Vivanco, "Regulation and Function of Root Exudates," *Plant, Cell & Environment* 32, no. 6 (June 2009): 666–81, https://doi.org/10.1111/j.1365-3040.2009.01926.x; D. L. Jones, C. Nguyen, and R. D. Finlay, "Carbon Flow in the Rhizosphere: Carbon Trading at the Soil-Root Interface," *Plant and Soil* 321 (February 2009): 5–33, https://doi.org/10.1007/s11104-009-9925-0; Shulbhi Verma and Amit Verma, "Plant Root Exudate Analysis," in *Phytomicrobiome Interactions and Sustainable Agriculture*, ed. Amit Verma et al. (Hoboken, NJ: Wiley, 2021), 1–14.

22. Leu, "Commentary V."
23. Leu, *Growing Life.*
24. Andreas Gattinger et al., "Enhanced Top Soil Carbon Stocks Under Organic Farming," *PNAS* 109, no. 44 (October 2012): 18226–31, https://doi.org/10.1073/pnas.1209429109; Eduardo Aguilera et al., "Managing Soil Carbon for Climate Change Mitigation and Adaptation in Mediterranean Cropping Systems: A Meta-Analysis," *Agriculture, Ecosystems & Environment* 168 (March 2013), 25–36, https://doi.org/10.1016/j.agee.2013.02.003; Leu, "Commentary V"; Teague et al., "The Role of Ruminants."
25. S. A. Khan et al., "The Myth of Nitrogen Fertilization for Soil Carbon Sequestration," *Journal of Environmental Quality* 36, no. 6 (November 2007): 1821–32, https://doi.org/10.2134/jeq2007.0099; R. L. Mulvaney, S. A. Khan, and T. R. Ellsworth, "Synthetic Nitrogen Fertilizers Deplete Soil Nitrogen: A Global Dilemma for Sustainable Cereal Production," *Journal of Environmental Quality* 38, no. 6 (October 2009): 2295–2314, https://doi.org/10.2134/jeq2008.0527; Meiling Man et al., "Altered Soil Organic Matter Composition and Degradation After a Decade of Nitrogen Fertilization in a Temperate Agroecosystem," *Agriculture, Ecosystems & Environment* 310 (April 2021): 107305, https://doi.org/10.1016/j.agee.2021.107305.
26. R. Lal, "Sequestration of Atmospheric CO_2 in Global Carbon Pools," *Energy and Environmental Science* 1 (July 2008): 86–100, https://doi.org/10.1039/B809492F.
27. Prescott et al., "Managing Plant Surplus Carbon."
28. Verma and Verma, "Plant Root Exudate Analysis."
29. Sheila Christopher, Rattan Lal, and Umakant Mishra, "Regional Study of No-Till Effects on Carbon Sequestration in Midwestern United States," *Soil Science Society of*

America Journal 73, no.1 (January 2009): 207–16, https://doi.org/10.2136/sssaj2007.0336; Verma and Verma, "Plant Root Exudate Analysis"; Leu, "Commentary V"; Leu, *Growing Life*.

30. Leu, *Growing Life*.
31. R. Lal, "Soil Carbon Sequestration Impacts on Global Climate Change and Food Security," *Science* 304, no. 5677 (June 2004): 1623–27, https://doi.org/10.1126/science.1097396; R. Lal et al., "Soil Carbon Sequestration to Mitigate Climate Change and Advance Food Security," *Soil Science* 172, no. 12 (December 2007): 943–56, https://doi.org/10.1097/ss.0b013e31815cc498; Shu Kee Lam et al., "The Potential for Carbon Sequestration in Australian Agricultural Soils is Technically and Economically Limited," *Scientific Reports* 3 (July 2013): 2179, https://doi.org/10.1038/srep02179; Jan Willem van Groenigen et al., "Sequestering Soil Organic Carbon: A Nitrogen Dilemma," *Environmental Science & Technology* 51, no. 9 (April 2017): 4738–39, https://doi.org/10.1021/acs.est.7b01427; Robert E. White, "The Role of Soil Carbon Sequestration as a Climate Change Mitigation Strategy: An Australian Case Study," *Soil Systems* 6, no. 2 (May 2022): 46, https://doi.org/10.3390/soilsystems6020046.
32. UN Convention to Combat Desertification, "Global Land Outlook," 2022.
33. Fred Pearce, "Rivers in the Sky: How Deforestation Is Affecting Global Water Cycles," *Yale Environment 360*, July 24, 2018, https://e360.yale.edu/features/how-deforestation-affecting-global-water-cycles-climate-change.
34. George A. Ban-Weiss et al., "Climate Forcing and Response to Idealized Changes in Surface Latent and

Sensible Heat," *Environmental Research Letters* 6, no. 3 (September 2011): 034032, https://doi.org/10.1088/1748-9326/6/3/034032.

35. Mallory L. Barnes et al., "A Century of Reforestation Reduced Anthropogenic Warming in the Eastern United States," *Earth's Future* 12, no. 2 (February 2024): e2023EF003663, https://doi.org/10.1029/2023EF003663.
36. Jean-Francois Bastin et al., "The Global Tree Restoration Potential," *Science* 365 no. 6448 (July 2019): 76–79, https://doi.org/10.1126/science.aax0848.
37. IPCC Working Group III, "Summary for Policymakers," 2014.

Chapter 4. Success Stories

1. "Scientists Measure Severity of Drought During the Maya Collapse," University of Cambridge, August 2, 2018, https://www.cam.ac.uk/research/news/scientists-measure-severity-of-drought-during-the-maya-collapse.
2. Richardson Gill, *The Great Maya Droughts: Water, Life, and Death* (Albuquerque: University of New Mexico Press, 2018).
3. Alison Kyra Carter, "Angkor Wat Archaeological Digs Yield New Clues to Its Civilization's Decline," *The Conversation*, June 3, 2019, https://theconversation.com/angkor-wat-archaeological-digs-yield-new-clues-to-its-civilizations-decline-116793.
4. Dr. Vandana Shiva, Dr. Mira Shiva, and Dr. Vaibhav Singh, *Poisons in Our Food* (Dehradun, India: Natraj Publishers, 2012).
5. Dr. Mae-Wan Ho, "Farmer Suicides and Bt Cotton Nightmare Unfolding in India," Science in Society Archive, June 1, 2010, http://www.i-sis.org.uk/farmersSuicidesBtCottonIndia.php.

6. Dr. Vandana Shiva and Dr. Vaibhav Singh, *Health Per Acre: Organic Solutions to Hunger & Malnutrition* (Rome, Italy: Navdanya International, 2017), https://navdanya international.org/wp-content/uploads/2017/07/Health -Per-Acre.pdf.
7. "Loess Plateau Watershed Rehabilitation Project," Environmental Education Media Project, August 5, 2013, https:// eempc.org/loess-plateau-watershed-rehabilitation-project.
8. "Restoring China's Loess Plateau," The World Bank, March 15, 2007, https://www.worldbank.org/en/news /feature/2007/03/15/restoring-chinas-loess-plateau.

Chapter 5. Business as Usual Is Not an Option

1. UN Convention to Combat Desertification, "Global Land Outlook," 2022.
2. Lal, "Soil Carbon Sequestration Impacts."
3. Kenneth Skrable, George Chabot, and Clayton French, "World Atmospheric CO_2, Its 14C Specific Activity, Non-fossil Component, Anthropogenic Fossil Component, and Emissions (1750–2018)," *Health Physics* 122, no. 2 (February 2022): 291–305, https://doi.org/10.1097 /HP.0000000000001485.
4. Leu, *Growing Life*.
5. García-García et al., "Soil Heat Extremes."
6. Prescott et al., "Managing Plant Surplus Carbon"; Verma and Verma, "Plant Root Exudate Analysis."
7. Sir Paweł Edmund de Strzelecki, *Physical Description of New South Wales and Van Diemen's Land: Accompanied by a Geological Map, Sections and Diagrams, and Figures of the Organic Remains* (London: Longman, Brown, Green and Longmans, 1845).
8. Strzelecki, *Physical Description of New South Wales*.
9. García-García et al., "Soil Heat Extremes."

10. Christine Jones, "Submission to the Environment and Natural Resources Committee: Inquiry into Soil Sequestration in Victoria," December 17, 2009, https://www.amazingcarbon.com/PDF/JONES-SoilSequestrationInquiry(17Dec09).pdf.
11. André Leu, *Poisoning Our Children* (Acres U.S.A., 2018).
12. "Farming Systems Trial," Rodale Institute, https://rodaleinstitute.org/science/farming-systems-trial.
13. Terry Cacek and Linda L. Langner, "The Economic Implications of Organic Farming," *American Journal of Alternative Agriculture* 1, no.1 (Winter 1986): 25–29, https://doi.org/10.1017/S0889189300000758.
14. "Wake Up Before It Is Too Late, Make Agriculture Truly Sustainable Now For Food Security in a Changing Climate," Trade and Environment Review, United Nations Conference on Trade and Development, 2013, https://unctad.org/publication/trade-and-environment-review-2013; N. Nemes, "Comparative Analysis of Organic and Non-Organic Farming Systems: A Critical Assessment of Farm Profitability," Food and Agriculture Organization of the United Nations, 2009, https://www.fao.org/documents/card/en/c/28b2b959-80f1-5244-bd52-0a98806d37ea.
15. "Organic Agriculture and Food Security in Africa," UNEP-UNCTAD, September 2008, https://unctad.org/system/files/official-document/ditcted200715_en.pdf.
16. K. Delate and C. A. Cambardella, "Organic Production: Agroecosystem Performance During Transition to Certified Organic Grain Production," *Agronomy Journal* 96 (2004): 1288–98, http://naldc.nal.usda.gov/download/8283/PDF.
17. D. W. Lotter, R. Seidel, and W. Liebhardt, "The Performance of Organic and Conventional Cropping Systems

in an Extreme Climate Year," *American Journal of Alternative Agriculture* 18, no. 3 (September 2003): 146–54, https://doi.org/10.1079/AJAA200345; David Pimentel, Rodolfo Zuniga, and Doug Morrison, "Update on the Environmental and Economic Costs Associated with Alien-Invasive Species in the United States," *Ecological Economics* 53, no. 3 (February 2005): 273–88, https://doi.org/10.1016/j.ecolecon.2004.10.002.

18. Lotter, Seidel, and Liebhardt, "The Performance of Organic."
19. Lotter, Seidel, and Liebhardt, "The Performance of Organic."
20. John P. Reganold, Lloyd F. Elliott, and Yvonne L. Unger, "Long-Term Effects of Organic and Conventional Farming on Soil Erosion," *Nature* 330 (November 1987): 370–72, https://doi.org/10.1038/330370a0.
21. Reganold, Elliott, and Unger, "Long-Term Effects of Organic."
22. L. E. Drinkwater, P. Wagoner, and M. Sarrantonio, "Legume-Based Cropping Systems Have Reduced Carbon and Nitrogen Losses," *Nature* 396 (November 1998): 262–65, https://doi.org/10.1038/24376; Rick Welsh, *The Economics of Organic Grain and Soybean Production in the Midwestern United States*, Policy Studies Report No. 13 (Greenbelt, MD: H. A. Wallace Institute for Alternative Agriculture, 1999); Pimentel, Zuniga, and Morrison, "Update on the Environmental"; Rodale Institute, "Farming Systems Trial."
23. Joshua L. Posner, Jon O. Baldock, and Janet L. Hedtcke, "Organic and Conventional Production Systems in the Wisconsin Integrated Cropping Systems Trials: I. Productivity 1990–2002," *Agronomy Journal* 100, no. 2 (March 2008): 253–60, https://doi.org/10.2134/agronj2007.0058.
24. Rodale Institute, "Farming Systems Trial."

Chapter 6. Agave Power

1. Nicoletta Maestri, "The History and Domestication of Agave," *Thought Co.*, January 21, 2020, https://www.thoughtco.com/domestication-history-of-agave-americana-169410.
2. Park S. Nobel, *Desert Wisdom/Agaves and Cacti: CO_2, Water, Climate Change* (iUniverse: 2009), 132.

Chapter 7. Scaling Up

1. Claire Marshall, "Kenya's Ogiek People Being Evicted for Carbon Credits - Lawyers," *BBC News*, November 9, 2023, https://www.bbc.com/news/world-africa-67352067; Blue Carbon, "Government of Kenya Accelerates Towards Compliance Market with UAE's Blue Carbon under Article 6," Media Release, October 25, 2023, https://www.newsfilecorp.com/release/185173.
2. Harry W. Fischer et al., "Community Forest Governance and Synergies Among Carbon, Biodiversity and Livelihoods," *Nature Climate Change* 13 (November 2023): 1340–47, https://doi.org/10.1038/s41558-023-01863-6.
3. Alan Buis, "The Atmosphere: Getting a Handle on Carbon Dioxide," NASA, October 9, 2019, https://climate.nasa.gov/news/2915/the-atmosphere-getting-a-handle-on-carbon-dioxide/; Lindsey and Dahlman, "Climate Change."

Appendix. Calculations

1. Schmidt et al., "Attribution of the Present-Day."
2. Megan B. Machmuller et al., "Emerging Land Use Practices Rapidly Increase Soil Organic Matter," *Nature* Communications 6 (April 2015): 6995, https://doi.org/10.1038/ncomms7995.
3. Jones, "Inquiry into Soil Sequestration in Victoria."

Index

Index

Index

J

K

L

About the Authors

Ronnie Cummins was the cofounder and director of the Organic Consumers Association (OCA), a nonprofit, US-based network of more than two million consumers dedicated to safeguarding organic standards and promoting a healthy, just, and regenerative system of food, farming, and commerce. Cummins also served on the steering committee of Regeneration International and OCA's Mexican affiliate, Vía Orgánica. He was the coauthor of *The Truth About COVID-19* and author of *Grassroots Rising* and *Genetically Engineered Food.*

Julia Leu

André Leu is the International Director of Regeneration International, an organization he cofounded in 2015 with Dr. Vandana Shiva, Ronnie Cummins, Dr. Hans Herren, and Steve Rye, and which has now grown to 570 partners in 75 countries, advancing projects in agroecology, permaculture, AMP grazing, agroforestry, and biological, organic, and ecological agriculture. Dr. Leu holds a Doctorate of Science in Environmental and Agricultural Systems and is an Adjunct Professor of regenerative agriculture at South Seas University. His books include *Growing Life*, *Poisoning Our Children*, and *The Myths of Safe Pesticides*. Dr. Leu and his wife, Julia, live on their organic tropical fruit farm in Daintree, Australia.